Reviews and critical articles covering the entire field of normal anatomy (cytology, histology, cyto- and histochemistry, electron microscopy, macroscopy, experimental morphology and embryology and comparative anatomy) are published in Advances in Anatomy, Embryology and Cell Biology. Papers dealing with anthropology and clinical morphology that aim to encourage cooperation between anatomy and related disciplines will also be accepted. Papers are normally commissioned. Original papers and communications may be submitted and will be considered for publication provided they meet the requirements of a review article and thus fit into the scope of "Advances". English language is preferred, but in exceptional cases French or German papers will be accepted.

It is a fundamental condition that submitted manuscripts have not been and will not simultaneously be submitted or published elsewhere. With the acceptance of a manuscript for publication, the publisher acquires full and exclusive copyright for all languages and countries.

Twenty-five copies of each paper are supplied free of charge.

Manuscripts should be addressed to

Prof. Dr. F. **BECK**, Howard Florey Institute, University of Melbourne, Parkville, 3000 Melbourne, Victoria, Australia

Prof. Dr. B. **CHRIST**, Anatomisches Institut der Universität Freiburg, Abteilung Anatomie II, Albertstr. 17, D-79104 Freiburg, Germany

Prof. Dr. W. **KRIZ**, Anatomisches Institut der Universität Heidelberg, Im Neuenheimer Feld 307, D-69120 Heidelberg, Germany

Prof. Dr. W. **KUMMER**, Institut für Anatomie und Zellbiologie, Universität Gießen, Aulweg 123, D-35385 Gießen, Germany

Prof. Dr. E. **MARANI**, Leiden University, Department of Physiology, Neuroregulation Group, P.O. Box 9604, 2300 RC Leiden, The Netherlands

Prof. Dr. R. **PUTZ**, Anatomische Anstalt der Universität München, Lehrstuhl Anatomie I, Pettenkoferstr. 11, D-80336 München, Germany

Prof. Dr. Dr. h.c. Y. **SANO**, Department of Anatomy, Kyoto Prefectural University of Medicine, Kawaramachi-Hirokoji, 602 Kyoto, Japan

Prof. Dr. Dr. h.c. T. H. **SCHIEBLER**, Anatomisches Institut der Universität, Koellikerstraße 6, D-97070 Würzburg, Germany

Ph. D. Gary C. **SCHOENWOLF**, Department of Neurobiology and Anatomy, University of Utah School of Medicine, 50 N. Medical Drive, Salt Lake City, Utah 84132, USA

Prof. Dr. K. **ZILLES**, Universität Düsseldorf, Medizinische Einrichtungen, C. u. O. Vogt-Institut, Postfach 101007, D-40001 Düsseldorf, Germany

Advances in Anatomy
Embryology and Cell Biology

Vol. 158

Springer-Verlag
Berlin Heidelberg
GmbH

Nadya Stefanova
Wladimir Ovtscharoff

Sexual Dimorphism of the Bed Nucleus of the Stria Terminalis and the Amygdala

With 44 Figures and 24 Tables

Springer

Nadya Stefanova, MD, PhD
Wladimir Ovtscharoff, MD, PhD, DSc
Departmen of Anatomy and Histology
Medical University Sofia
1 G. Sofiisky St.
1431 Sofia, Bulgaria

E-mail: nadya-stefanova@rocketmail.com
E-mail: ovcharov@medfac.acad.bg

ISSN 0301-5556
ISBN 978-3-540-67683-6

Library of Congress-Cataloging-in-Publication-Data

Stefanova N. (Nadya), 1969-
Sexual dimorphism of the bed nucleus of the stria terminalis and the amygdala/N.
Stefanova, W. Ovtscharoff
p. cm. — (Advances in anatomy, embryology, and cell biology ; v. 158)
Includes bibliographical references and index.
ISBN 978-3-540-67683-6   ISBN 978-3-642-57269-2 (eBook)
DOI 10.1007/978-3-642-57269-2
1. Developmental neurobiology. 2. Brain—Sex differences. 3. Amygdaloid body. I.
Ovtscharoff, V. II. Title.III. Series.

© Springer-Verlag Berlin Heidelberg 2000
Originally published by Springer-Verlag Berlin Heidelberg New York in 2000

Production: PRO EDIT GmbH, 69126 Heidelberg, Germany
Printed on acid-free paper – SPIN: 10718118   27/3136wg - 5 4 3 2 1 0

Preface

Sex differences are observed in various physiological, behavioral, and psychic functions, including reproductive behavior, aggression, emotions, and cognition. Such differences are expressed even in early childhood with preferences to definite activities. It has been generally accepted that differences between genders are formed under the influence of biological as well as environmental factors. The existence of sex differences in functions of the central nervous system has suggested that there are also morphological sex differences.

In recent years several reports on sexual dimorphism in the brain of vertebrates have been published. However, the mechanisms of sexual differentiation of the central nervous system remain unclear in most cases. It is often difficult to correlate morphological sex differences to differences in definite function or behavior.

We set out to explore the sexual dimorphism of the limbic system and especially the bed nucleus of the stria terminalis and the amygdala, which are considered generally to be occupied with the control of reproductive behavior and autonomic and complicated psychic functions. Several reports on sexual dimorphism of these structures have been published. Some of them have been directed to the total neuronal number and the volume of the nuclei, while others have concentrated on definite subpopulations of neurons. In many cases the mechanisms of sexual differentiation were tested, but sometimes they could not be established. Several attempts have been made to elucidate the functional significance of sexual dimorphism of these limbic structures, yet it remains unknown for most studies carried out.

The present study confirms the previously reported existence of sex differences in the total neuronal numbers and volumes of the bed nucleus of the stria terminalis and the amygdala, as well as the mechanisms of their generation. Additionally, it provides new evidence for the sexual dimorphism and differentiation of GABAergic, leucine-enkephalin-containing and parvalbumin-immunoreactive neurons in the bed nucleus of the stria terminalis and amygdala of the rat. Together with testing the gender differences, the use of different age groups made it possible to follow the changes in numbers of neuronal subpopulations, mentioned above, during ageing. Thus, our study contributes with new results to the understanding of the sexual differentiation of the bed nucleus of the stria terminalis and the amygdala. It gives some new insight into the problem of setting a rule for sexual differentiation of the brain. Our results provide morphological and immunocytochemical data that may be used for further studies on sexually dimorphic circuitry and its functional significance. Our study could also be important for understanding sex differences in the pathology of the central nervous system.

Acknowledgements

This work was performed in the Department of Anatomy and Histology, Medical University, Sofia, Bulgaria. It was accomplished with the technical assistance of Mrs. S. Ilieva, Mrs. K. Zlatanova, and Mrs. D. Lasheva. We highly appreciate the valuable help of Mrs. D. Brazitsova with morphometric analysis of the data. Assoc. Prof. A. Bozhilova-Pastirova is greatly acknowledged for the continuous support and creative ideas that made it possible to fulfil this study. All the antibodies applied are gifts by Prof. E. Marani from the Neuroregulation Group of the Department of Physiology, Leiden University, the Netherlands and Prof. C. Pilgrim, Department of Anatomy and Cell Biology, University of Ulm, Germany.

Contents

1 Introduction

Sex differences are present in the endocrine functions and behavior of all vertebrates. These differences are especially well delineated in the reproductive behavior of both genders. It has been suggested that functional differences between males and females are connected with sexual dimorphism in the morphology of the central nervous system (CNS). During recent decades, studies in this field revealed significant sex differences in various structures of the CNS. Among them are structures that are directly related to the control of reproductive functions, as well as grey and white matter elements, which take part in the control of other functions.

1.1
Sexual Differentiation of the Brain

The dependence of adult reproductive behavior on the presence of sex steroids in the perinatal period strongly supported the idea of looking for morphological sex differences in the brain. Pfeiffer's experiments (1936) have shown that the expression of male-type gonadotropin secretion in adult male rats is dependent on testicular factors, secreted in early postnatal life. Male-type gonadotropin secretion could be established in females, if testis transplantation is performed soon after birth. At the same time, castration of newborn males leads to female-type gonadotropin secretion and behavioral characteristics in adulthood (Barraclough and Gorski 1962). In 1959, Phoenix et al. found that female guinea pigs, which had been exposed to high levels of testosterone before birth, showed disturbed reproductive behavior in adulthood. Phoenix et al. (1959) suggested the possibility that androgens have an organizational effect on developing brain, and disturbances in reproductive behavior are a reflection of the reorganization of the brain structure. With the accumulation of data on the dependence of sex-specific behaviors on the presence of sex hormones in the perinatal period, the idea of sexual differentiation of the brain is accepted.

According to the classical theory for sexual differentiation of the brain (see Stefanova et al. 1996), if testosterone is present during "critical periods" of early development, masculinization and defeminization in behavior and morphology of the CNS take place. These processes are irrespective of genetic sex. They are irreversible organizational effects. Testosterone under the influence of the enzyme cytochrome P450 aromatase (Naftolin et al. 1975; McEwen et al. 1977) in neurons is converted to estradiol and thus fulfils its role (Fig. 1).

Therefore, the influence of the female sex steroid estradiol is directly responsible for the masculinization and defeminization of brain structures. In females this does

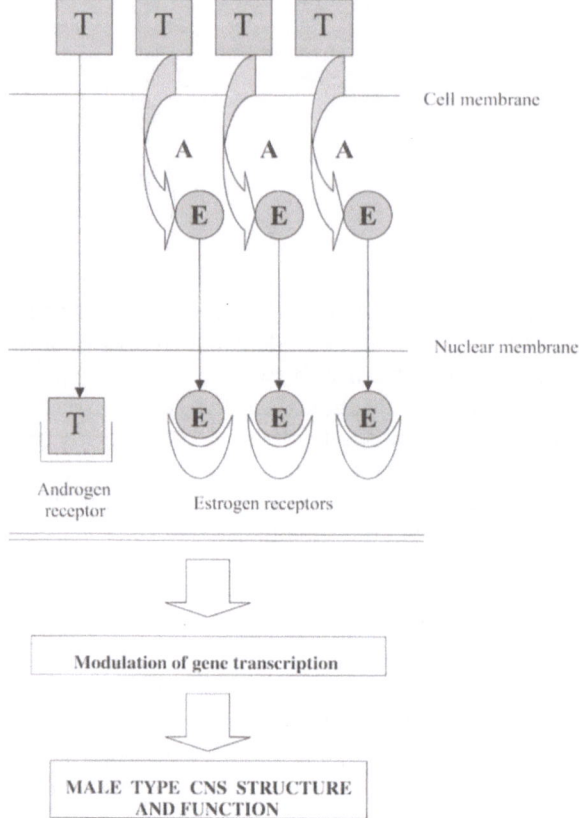

Fig. 1. Summarized scheme of the classical theory for sexual differentiation of the brain under the influence of testosterone (*T*), which is converted by the enzyme aromatase (*A*) into estradiol (*E*) during "critical periods" of development

not happen, because estrogen-binding alpha-fetoprotein, which is synthesized by fetal liver, prevents binding of estrogens to estrogen receptors. It is thus accepted that estrogens in female fetuses cannot influence the neural genome by acting on their receptors. Organizational effects of sex hormones during the critical period of sexual differentiation of the brain are permanent and irreversible. Together with this, sex steroids also perform activational effects on the brain structure during later stages of ontogenesis, but these effects are reversible (Beatty 1979). The critical period for sexual differentiation of rodent brain lasts from the last days of fetal life till 7–10 days after birth. Rat testes begin synthesizing androgens at day 13 of intrauterine development and testosterone is present till the tenth postnatal day (Kelly 1991). According to Weisz and Ward (1980), in male rats, testosterone levels increase significantly between postconception days 17 and 18 and reach a peak on days 18 and 19. Plasma testosterone is significantly higher in males than in females on each day from the 18th gestation day till the fifth postnatal day. Rhees et al. (1990a,b) have determined that the critical period of sexual differentiation for the sexually dimorphic nucleus of the preoptic area (SDN-POA) in rats begins on the 18th intrauterine day and terminates on the fifth postnatal day. It is considered that there is a specific term for sexual differentiation of each brain structure, which depends on regional hormone sensitivity and hormonal dose (Gorski 1985a,b). All these data have reasoned the acceptance of the epigenetic character of brain sexual differentiation irrespective of genetic sex.

The presence of estrogen and androgen receptors in the developing brain is proof that gonadal steroids act directly on nervous tissue. They are representatives of ligand-activated nuclear transcription factors. In most cases these receptors are found in the nuclei of target cells (Kawata 1995). When the gonadal steroid binds to its receptor it affects the genome and thus results in morphological or functional cellular reply. Recently, it has been proven that sex steroids can act also on the surface of the neuron and change ion permeability and secretion of neurohormones and neuro-transmitters (McEwen 1991).

Two types of estrogen receptors have been described – ERα and ERβ (Kuiper et al. 1996; Shughrue et al. 1997a,b). Both types of estrogen receptors are expressed in the brain. However, a certain regional specificity of their distribution is observed (Pfaff and Keiner 1973; Koch and Ehret 1989). Estrogen receptors are dense in the hypothalamus, the amygdaloid complex, the bed nucleus of the stria terminalis, and the septum. Levels of estrogen receptors change with age in a region-specific manner (Vito and Fox 1982; O'Keefe and Handa 1990; Yokosuka et al. 1995). Distribution of estrogen receptors in the brain shows sex differences in certain areas (Kühneman et al. 1995). The expression of estrogen receptors is sensitive to testosterone levels in the critical period of sexual differentiation (Don Carlos et al. 1995). Studies which describe the distribution of androgen receptors in the brain (Barley et al. 1975; Choate and Resko 1992; Clancy et al. 1992, 1994; Sar et al. 1990; Iqbal et al. 1995) also establish dependence of their levels on testosterone.

With the aid of in situ hybridization, Simerly et al. (1990) presented the distribution of cells containing estrogen and androgen receptor mRNA in the brain of male and female rats. They concluded that estrogen receptors are mostly expressed in regions occupied with the control of gonadotropin secretion, while in the same regions, androgen receptors are few. At the same time, they found an overlapping of estrogen and androgen receptor distribution in the amygdala, the bed nucleus of the stria terminalis, the medial preoptic area, arcuate nucleus, periventricular nucleus, ventromedial nucleus, and central grey. Whether one and the same neuron could express both types of receptors remained unclear.

The distribution of androgen and estrogen receptors was found to be quite similar to that of aromatase-containing neurons (Roselli and Resko 1993; Shinoda 1994; Lephart 1996). As mentioned above, aromatase is the enzyme responsible for the conversion of testosterone to estradiol in the neuron and realizing androgen effects of sexual differentiation of the brain (Naftolin et al. 1975). Using enzyme activity (Lephart et al. 1992a,b) and immunohistochemistry (Jakab et al. 1993), it is established that the highest levels of aromatase are present in the hypothalamus and limbic structures, especially the bed nucleus of the stria terminalis, the medial and cortical amygdaloid nuclei, ventromedial hypothalamus, and anterior hypothalamus. Aromatase activity is regionally specific and changes in different periods of individual life. It is highest during the critical period of sexual differentiation of the brain (Lephart et al. 1992a; Raab et al. 1995a). Aromatase expression is sexually different in certain brain regions (Roselli and Resko 1993; Beyer et al. 1993; Hutchison et al. 1995), which could be the main reason for differences in estrogen levels in these regions during the critical period of sexual differentiation. Several factors can influence aromatase expression in the brain. Among these are androgens (Roselli and Resko 1993), norepinephrin and cyclic AMP (Lephart et al. 1992 b), prenatal stress (Weisz et al. 1982), ethanol (Kelce et al. 1980), and nicotine (Van Ziegler et al. 1991). These data confirm the complicated and still unclear mechanisms of control of aromatase in the CNS and thus its importance for the formation of sex differences in brain structure and function.

Although the epigenetic effect of gonadal steroids on sexual differentiation of the brain is widely accepted, it is suggested that genetic factors may also contribute to the regulation of this process. Results from sex-specific embryonic cell cultures (Reisert and Pilgrim 1991, 1995) confirm that dopaminergic mesencephalic neurons can develop their morphological and functional gender characteristics in the absence of sex steroids. Lahr et al. (1995) found that the Sry gene, which is responsible for the differentiation of testes, is also transcribed in the hypothalamus and mesencephalon of adult male, but not female rats. Several genetic mechanisms are discussed, proving that autosome and gonosome genes could affect sex-specific functions, behaviors, and morphological characteristics of the brain (see Maxon 1997).

Recent studies discuss much richer interactions of factors and mechanisms that make it difficult to predict their contribution to sexual differentiation in diverse brain regions (Breedlove et al. 1999; Jordan 1999).

Our investigation is concentrated on two limbic structures – the bed nucleus of the stria terminalis (BST) and the amygdala. Several reasons have determined our choice: (a) the BST and the amygdala are characterized by a high concentration of estrogen and androgen receptors, as well as high aromatase expression; (b) up to now, data prove the existence of sex differences in the structure of the BST and the amygdala; (c) the BST and the amygdala are connected between themselves and with other sexually dimorphic nuclei in a sexually dimorphic circuitry, which participates in the control of reproductive behaviors in males and females.

1.2
Bed Nucleus of the Stria Terminalis

The BST is a cerebral neuronal group, closely related to the stria terminalis. According to the classical description of Johnson (1923), the BST is a "band" of grey matter that accompanies the stria terminalis and widens at its rostral and caudal ends. Young (1936) has named the rostral part of the BST nucleus interstitialis striae terminalis. A small number of cells scattered along the fibers of the stria terminalis form the intra-amygdaloid continuum of the nucleus caudally. In rodents, this caudal part is not well developed, and in the literature the region immediately near the anterior commissure is usually described as the BST or nucleus interstitialis striae terminalis.

Traditionally, the BST is divided into two main parts – medial and lateral (Krettek and Price 1978a). Andy and Stephan (1968) describe three parts of the nucleus in humans – anterior, external, and internal. On the basis of immunohistochemical and projection characteristics, Moga et al. (1989) supplement the description of De Olmos et al. (1985) and Paxinos and Watson (1986) and additionally subdivide the parts of

AFFERENTS   **EFFERENTS**

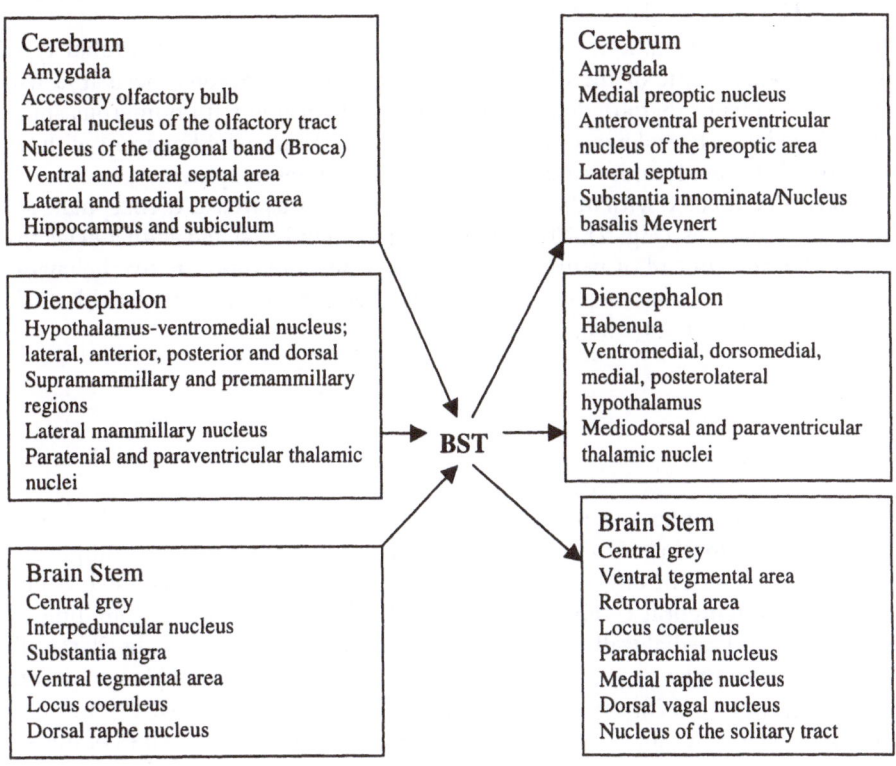

Fig. 2. Summarized scheme of the main afferent (Weller and Smith 1982; Vankova 1991; Petit et al. 1995; Canteras et al. 1992; Canteras and Swanson 1992; Simerly and Swanson 1988; Morin et al. 1994; Bressler and Baum 1996) and efferent (Gray and Magnuson 1987; Behzadi et al. 1990; Holstege et al. 1985; Moga et al. 1989; Loughlin and Fallon 1985; Numan and Numan 1997; McLean et al. 1983; Churchill et al. 1996; Gritti et al. 1994, 1998; Otake and Nakamura 1995; Otake et al. 1995; Berk and Finkelstein 1981; Cullinan et al. 1993; Herman and Cullinan 1997; Simerly and Swanson 1986; Hutton et al. 1998; Caffe et al. 1987; Fuller et al. 1987; Vankova 1991) connections of the bed nucleus of the stria terminalis (*BST*)

the BST. They determine a medial, lateral, and ventral part of the rat BST. The medial part is spread between the dorsomedial border of the nucleus accumbens and rostral thalamus. It consists of six subnuclei – anterior medial, ventral medial, posterior medial, posterior intermediate, subventricular, and intermedial. The lateral part of the BST is also divided into six subnuclei – anterior, dorsal, ventral, posterior, supracapsular, and juxtacapsular. The ventral part of the BST includes the ventral, lateral, and preoptic subnucleus, as well as the parastrial nucleus. The cytoarchitecture of the rat BST has also been extensively described by Ju and Swanson (1989), but they use a rostrocaudal approach and divide the BST into an anterior and posterior part. According to De Olmos (1990), the human BST includes medial and lateral as well as anterior and posterior parts. A comparison of the most widely spread descriptions of the BST is presented by Moga et al. (1989).

Using Golgi impregnation, McDonald (1983) has described the neuronal types which are characteristic of the different BST parts. Ovoid neuronal perikarya with four to five dendrites that are densely branching and covered with numerous spines are described in the lateral part of the BST. In the medial part, ovoid neurons with two to three dendrites are found, weakly branching and with fewer spines.

The BST is connected with other parts of the cerebrum, as well as diencephalic and brain stem structures (Fig. 2). The medial part is mainly related with nuclei that participate in the control of neuroendocrine functions and reproductive behaviors. On the other hand, the lateral part of the BST has reciprocal connections with nuclei of central autonomic regulation.

1.3
The Amygdala

The amygdaloid complex is a major element of the limbic system. It is related with modulation of endocrine responses, visceral effector mechanisms, and complicated behavioral reactions. The amygdala is described as a neuronal group in the dorsome-dial part of the temporal cerebral lobe. The amygdala participates in building the ventral, upper, and medial walls of the inferior horn of the lateral ventricle. Classically, the amygdaloid complex is divided into two main nuclear groups – corticomedial and basolateral (Johnson 1923; Brockhaus 1938; Gloor 1960). Several descriptions of the nuclear subdivisions of the amygdaloid complex exist. During early stages of investigation scientists usually based their results on the outlook of the nuclei, not taking into consideration the afferent and efferent connections and chemoarchitecture, which was the reason for several discrepancies (Gloor 1960; Hall 1972; De Olmos 1990). Krettek and Price (1978a) divide the amygdala into deep amygdaloid nuclei, cortical structures connected with the amygdala and other cell masses, connected with the amygdala. The basolateral group comprises the lateral, basolateral, basome-dial, and central nucleus. The group of cortical structures connected with the amygdala comprises the medial nucleus, the amygdalo-hippocampal area, anterior and posterior cortical nuclei, periamygdaloid cortex, and the nucleus of the lateral olfactory tract. Smaller cell groups are also included in the amygdaloid complex among its major parts. These are the bed nucleus of the accessory olfactory tract, intercalated neuronal masses, and intra-amygdaloid part of the BST. Krettek and Price (1978a) have given a successive description of neurons comprising each amygdaloid nucleus in the cat and rat in thionin-stained sections. In a series of studies, McDonald (1982a,b, 1984) presented descriptions of the amygdaloid nuclei applying Golgi impregnation. He confirmed the subdividing of the central nucleus into medial and lateral parts. The neurons in the lateral part possess numerous spines and resemble the medium cells of the striatum. In the medial part of the central nucleus there are mainly neurons with long, sparsely branching dendrites and a few spines, as well as cells with aspiny thick dendrites. Three types of neurons have been observed in the lateral and basolateral amygdaloid nucleus of the rat (McDonald 1982b, 1984) – spiny neurons, small ovoid neurons with spine-sparse dendrites, and small spherical neurons with short varicose dendrites. Cellular types have been described in the human basolateral amygdala too (Braak and Braak 1983).

The amygdaloid complex has a key role in the limbic system, sending intra-amygdaloid and intralimbic projections as well as projections to other cerebral, diencephalic, and brain stem nuclei (Fox 1949; Nauta 1961, 1962; Fallon and Moore 1978; Hopkins and Holstege 1978; Krettek and Price 1978b; Usunoff et al. 1979; Mehler 1980; Ottersen 1981, 1982; Russchen 1982a,b; Loughlin and Fallon 1983; Mizuno et al. 1985; Ragsdale Jr and Graybiel 1988; Yamano et al. 1988; Vankova 1991; Canteras et al. 1995; Da Costa Gomez and Behbehani 1995; Petrovich et al. 1996; Wright et al. 1996; McDonald and Mascagni 1997; Brodal 1998).

1.4
Chemoarchitecture of the BST and the Amygdala

Studies of the chemoarchitecture of the BST and the amygdala have proven that these limbic structures are quite rich in neurotransmitters and neuromodulators. Together with their various projections, they additionally complicate interactions with other brain structures and the intra-amygdaloid modulation of signals (Table 1).

1.5
Gender-Related Characteristics of the BST and the Amygdala

The BST and the amygdala are places of high concentration of estrogen and androgen receptors (Simerly et al. 1990) and high aromatase activity and expression (Shinoda et al. 1994; Wagner and Morrell 1997). A solid body of evidence proves the existence of sex differences in the morphology of the BST and the amygdala (discussed later).

1.6
Mediators of Reproductive Behavior

Sexual dimorphism of the transmitter systems in the CNS and the dependence on gonadal steroid levels suggest their contribution to the formation of sex-specific functions. It has been shown that cholecystokinin (CCK) participates in the regulation of female reproductive behavior (Mendelson and Gorzalka 1984), luteinizing hormone (LH) secretion (Kimura et al 1983; Hashimoto and Kimura 1986), and secretion of luteinizing hormone-releasing hormone (LHRH) (Micevych et al. 1986). Substance P takes part in the control of male reproductive behavior and secretion of LH and prolactin (Kostarczyk 1986). Vasopressin in the CNS also plays a role in the control of reproductive behavior and seasonal variations (De Vries 1990).

The inhibitory neurotransmitter gamma-aminobutyric acid (GABA) plays an important role in the central regulation of LHRH secretion. Direct synaptic contacts between GABAergic and LHRH neurons are established in the preoptic area (Horvath et al. 1993). Application of exogenous GABA and its agonists in the medial preoptic nucleus inhibits secretion of LH (Herbison et al. 1991a,b) and decreases gene expression of LHRH (Bergen et al. 1991). GABA inhibits female "lordosis" behavior through

Table 1. Some neurotransmitters in the bed nucleus of the stria terminalis (*BST*) and the amygdala

Neurotransmitters	BST	Amygdala
Peptides		
Vasoactive intestinal peptide (VIP)	Woodhams et al. 1983	Roberts et al. 1982; Cassell and Gray 1989
Neurotensin	Moga et al. 1989; Ju et al. 1989	Roberts et al. 1982; Cassell and Gray 1989
Substance P	Ju et al. 1989	Roberts et al. 1982; Neal Jr et al. 1989; Cassell and Gray 1989; Simerly et al. 1989
Cholecystokinin (CCK)	Ju et al. 1989	Roberts et al. 1982; ' Simerly et al. 1989
Neuropeptide Y	de Quidt et al. 1990; Burroughs et al. 1996	McDonald 1989; Burroughs et al. 1996
Galanin	Ju et al. 1989; Planas et al. 1994	Cassell and Gray 1989; Planas et al. 1994
Angiotensin	Lind and Ganten 1990	Lind and Ganten 1990
Corticotropin-releasing factor	Moga et al. 1989; Ju et al. 1989; Makino et al. 1994	Cassell and Gray 1989; Makino et al. 1994
Somatostatin	Woodhams et al. 1983	Roberts et al. 1982; Cassell and Gray 1989
Vasopressin	Planas et al. 1995	Caffe and Van Leeuwen 1983; Urban et al. 1990
Dinorphin	Neal Jr et al. 1989	Neal Jr et al. 1989
Enkephalin	Moga et al. 1989; Woodhams et al. 1983	Roberts et al. 1982; Cassell and Gray 1989
Monoamines		
Dopamine	Björklund and Lindvall 1984	Ben-Ari et al. 1975; Björklund and Lindvall 1984
Amino acids		
GABA	Mugnaini and Oertel 1985	Mugnaini and Oertel 1985

its action on GABA-A receptors in the preoptic area, while in the mediodorsal hypothalamus and the mesencephalon it has the opposite effect (McCarthy et al. 1991, 1995). It is considered that GABA plays a role in the formation of maternal behavior. Maternal behavior in lactating female rats is blocked by benzodiazepine antagonists, while benzodiazepines themselves stimulate this behavior in non-maternal animals (Hansen et al. 1985). Majewska et al. (1989) have shown that pregnancy-induced decrease in GABA-A receptors and increase in their affinity are connected with promotion of maternal behavior. The participation of the GABA-A-benzodiazepine-Cl– receptor complex in the development of maternal behavior has been proven by Del Cerro et al. (1995).

CCK, vasopressin, and substance P, which take part in the formation of sex-specific behaviors, have been found to have a sexually dimorphic expression in the BST and the amygdala (Van Leeuwen et al. 1985; Malsbury and McKay 1987; Micevych et al. 1988; De Vries and Al-Shamma 1990; Szot and Dorsa 1993; Lakhdar-Ghazal et al. 1995; Wang and De Vries 1995; Al Shamma and De Vries 1996). GABA distribution in the BST and in the amygdala is successively described (Mugnaini and Oertel 1985; Nitecka and Ben-Ari 1987) and yet no studies have been performed on its sexual dimorphism. Sex differences of GABA-immunoreactive neurons have been described in the rat striatum (Ovtscharoff et al. 1992, 1997; Bozhilova-Pastirova and Ovtscharoff 1996). Moreover, several studies have proven that GABA expression is under gonadal steroid control. Flügge et al. (1986) have found that GABAergic neurons in the medial preoptic area concentrate estrogen. Estrogen-induced morphological changes have been described in GABAergic neurons of the arcuate nucleus (Parduz et al. 1993). Exogenous estrogens increase GABA content in microdialysates from the medial preoptic nucleus (Demling et al. 1985; Herbison et al. 1991a,b). McCarthy et al. (1995) found that glutamate decarboxylase (GAD) mRNA in its two forms, GAD_{65} and $GAD_{67,}$ is modulated by estrogens.

Testosterone also has regulatory effects on GABAergic neurons. Grattan and Selmanoff (1993) have shown that castration of male rats causes an increase of steady-state GABA concentrations and a GABA turnover decrease in the diagonal band of Broca, the medial preoptic area, and the median eminence. GABA turnover increases in the medial septal nucleus and remains unaffected in the cortex, striatum, and hindbrain. Testosterone can abolish this effect in the rostral hypothalamus and the median eminence, whose GABAergic neurons exert negative feedback regulation on LHRH neurons (Grattan and Selmanoff 1994a,b).

GABA often coexists with other neuromodulators or transmitters. That is why GABAergic neurons can be divided into subpopulations. GABA has been shown to colocalize with somatostatin, neuropeptide Y, CCK, and vasoactive intestinal peptide in the basolateral amygdala (McDonald and Pearson 1989). Another group of neuropeptides that coexists with GABA are the Ca2+-binding proteins calbindin-D-28K, calretin, and parvalbumin.

Parvalbumin has been isolated from muscle tissue of lower vertebrates for the first time, and is functionally related to relaxation of fast-twitch muscle fibers through its binding to calcium (Heizmann 1984). Parvalbumin is present in other tissues including nervous tissue. It regulates the intracellular levels of calcium in neurons and thus participates in the control of various functions including neurotransmission. Parvalbumin is considered a functional buffer in the brain, but its exact role remains unclear. Distribution of parvalbumin in the brain is well documented (Celio 1990; Glezer et al.

1993; Pitkänen and Amaral 1993a,b; Conde et al. 1994; Wouterlood et al. 1995). Colocalization studies have revealed that parvalbumin-immunoreactive neurons usually form a subpopulation of GABAergic neurons (Cowan et al. 1990; Martinez-Cuijarro et al. 1994). In most cases these are interneurons, but some GABAergic projection neurons can also express parvalbumin (Freund 1989; Magnusson et al. 1996). Parvalbumin-immunoreactive neurons are expressed after the first postnatal week with different terms for various brain regions (Alcantara et al. 1993). This is probably related with the functional activity of different brain nuclei. Although data are present for the coexistence of parvalbumin and GABA and for the distribution of parvalbumin-immunoreactive neurons in the amygdala (Celio 1990; Sorvari et al. 1995), no studies on the sexual dimorphism of this neuronal subpopulation have been performed. Sex differences for parvalbumin-immunoreactive neurons have been reported in the rat striatum (Bozhilova-Pastirova and Ovtscharoff 1996; Kolev et al. 1997; Stoykov et al. 1998), nucleus accumbens (Karamfiloff et al. 1998), and sensorimotor cortex (Bozhilova-Pastirova et al. 1997b).

The opioidergic system also takes part in the modulation of reproductive behavior and in sex-steroid-dependent functions. Romano et al. (1990) have established that gene expression of proenkephalin in hypothalamic neurons is activated by estrogens. Enkephalins have been found to stimulate lordosis behavior in female rats. Van Furth et al. (1995) have also established the regulatory effect of the brain opioid system on male sexual behavior by stimulation of sexual motivation. It is considered that the corticomedial part of the amygdala is an element of the neural circuitry underlying motivation and processing of olfactory information with participation of beta-endorphin. Torii et al. (1996) have reported that injecting leucine-enkephalin in the third ventricle during estrogen priming inhibits lordosis behavior in female rats. Simerly et al. (1988) described sexual dimorphism of enkephalin-ergic neurons in the anteroventral periventricular nucleus of the preoptic area.

Present data suggest that the BST and the amygdala are places of high concentration of estrogen and androgen receptors. Both limbic structures show sexual dimorphism and participate in the regulation of reproductive behavior. On the other hand, GABA and endogenous opioids are also engaged in the control of reproductive behavior. Distribution of GABA and opioids in the BST and the amygdala is well documented, but there have been no studies performed on sexual dimorphism of these neurotransmitter systems and mechanisms of its generation.

2 Materials and Methods

2.1
Experimental Animals

Male and female Sprague-Dawley rats were used, maintained under a 12-h light:12-h dark cycle with free access to food and water. Rats were bred in an animal care facility at the Department of Anatomy and Histology, Medical University, Sofia.

Intact untreated animals were grouped into four age groups (Table 2):
1. 6 days old – early postnatal age before the end of the critical period of brain sexual differentiation.
2. 20 days old – prepubertal rats with accomplished critical period of brain sexual differentiation.
3. 3 months old – young postpubertal rats.
4. 1 year old – ageing rats.

Experimentally manipulated animals also comprised four groups (Table 2):
1. CN – male rats at 3 months of age which were castrated immediately after birth.
2. CP – male rats at 3 months of age which were castrated at puberty on postnatal day 35.
3. EA – male rats at 3 months of age which were treated with estrogen antagonist during the first 10 days of life.
4. AI – male rats at 3 months of age which were treated with aromatase inhibitor during the first 10 days of life.

Table 2. Number of male and female rats used in the study. Selected experiments for the morphometric study dispersed in four age and four experimental groups

	Total	6 days	20 days	3 months	1 year	CN	CP	EA	AI
Male	40	5	5	5	5	5	5	5	5
Female	20	5	5	5	5	–	–	–	–

CN, 3-month-old males, castrated on the first day of life; *CP*, 3-month-old males, castrated at puberty; *EI*, 3-month-old males, treated with estrogen antagonist in the first 10 days of life; *AI*, 3-month-old males, treated with aromatase inhibitor in the first 10 days of life.

2.2
Anesthesia

Intraperitoneal anesthesia with Thiopental or Nembutal at 40 mg/kg b. w. preceded all the perfusions. Castrations were done under ether anesthesia.

2.3
Castrations

After newborn male rats were deeply anesthetized, a 3–4-mm horizontal section between the anus and the urethra was performed. Testes were extirpated with the aid of fine forceps and the spermatic cord was cut. Blood loss was minimal due to the small size of the blood vessels and the mechanical hemostasia. The same operation was applied to pubertal male rats on their 35th day of life. Organ verification of the testes was always done under light microscope. All operated animals were left to survive till 3 months of age, before being sacrificed for this study.

2.4
Experimental Treatment

Newborn male rats were injected subcutaneously during the first 10 days of life with either: (1) the estrogen antagonist Tamoxifen (Generics Ltd., Potters Bar, Herts., UK) at a concentration of 0.02 mg diluted in sesame oil; (2) the aromatase inhibitor Lentaron (Ciba-Geigy Ltd., Basel, Switzerland) at a concentration of 0.005 mg diluted in saline. All the injected male rats were sacrificed at 3 months of age.

2.5
Fixation

After anesthesia, rats were perfused transcardially with 4% paraformaldehyde in 0.1 M phosphate buffer, pH 7.2–7.4. For the 6-day-old animals, 80–100 ml perfusion solution was used, and for all the other ages 150–200 ml. Each perfusion lasted about 15–20 min. Immediately after that, brains were removed from the skull, and were postfixed in the same fixative for 1 h at room temperature. After rinsing in phosphate buffer, the material was left overnight in buffer containing 15% sucrose at 4°C.

2.6
Sectioning

Coronal blocks with coordinates from +0.20 mm to –4.30 mm in reference to bregma (Paxinos and Watson 1986) including the BST and the amygdala were sectioned serially from rostral to caudal end on a freezing microtome (Reichert-Jung). Sections were 40 μm thick. Every fifth section was taken for the separate studies.

2.7
Nissl Staining

Section series for the Nissl staining were mounted on slides with 0.2% water solution of gelatin. After drying, slides were hydrated in descending alcohols (5 min in 96%, 80%, 70% ethanol), washed in distilled water and immersed for 30 min in 0.5% water solution of cresyl-violet. After staining, slides were shortly washed in water and differentiated in acidic alcohol (96% ethanol containing ice acetic acid). Sections were dehydrated in ascending alcohols (2–5 min in 70%, 80%, 2×96%, 2×100% ethanol) and immersed in xylol for 24 h. Sections were coverslipped with entellan (Merck).

2.8
Immunocytochemistry

The following specific antibodies were used: (a) anti-GABA (Sigma Bio Sciences, St. Louis, Missouri) polyclonal; (b) anti-leucine-enkephalin (Sigma Bio Sciences, St. Louis, Mo., USA) polyclonal; and (c) anti-parvalbumin (Sigma Bio Sciences) monoclonal.

The following second antibodies and marking substances were used: (a) goat anti-rabbit IgG, conjugated to 5 nm gold particles, whole molecule (Sigma Bio Sciences); (b) biotinylated anti-mouse IgG (Vector Laboratories Inc. Burlingame, Calif., USA); (c) silver enhancer IntenSETMBL (Amersham Life Sciences, Buckinghamshire, UK); (d) avidin-biotin peroxidase kit Vectastain ABC (Vector Laboratories Inc.); and (e) 3, 3-diaminobenzidine (DAB) (Sigma Bio Sciences).

The following normal serums were used: (a) normal goat serum (Sigma Bio Sciences); (b) bovine serum albumin (Sigma Bio Sciences).

The following buffers were used: (a) 0.1 M phosphate buffer (PB), pH 7.2–7.4; (b) 0.01 M phosphate buffer saline (PBS), pH 7.2–7.4; (c) 0.01 M Tris-buffer saline (TBS), pH 8.1; and (d) 0.05 M Tris-buffer-HCl (Tris-HCl), pH 7.6.

Sections that were processed for GABA and leucine-enkephalin immunocytochemistry were first incubated for 1 h in 5% normal goat serum, followed by incubation in the first antibody-anti-GABA (1:4000) or anti-leucine-enkephalin (1:1000) for 48 h. Rinsing in TBS preceded the incubation in 1% bovine serum albumin for 20 min. Incubation with the second antibody goat-anti-rabbit IgG, conjugated to 5 nm gold particles at 1:100, lasted for 1 h. A new series of rinsing was done first in TBS and then in distilled water. Silver intensification with the silver enhancement kit followed until visualization of the reaction occurred. A 2-min incubation in 2.5% sodium tiosulfate was followed by rinsing in TBS.

Sections for parvalbumin immunocytochemistry were incubated overnight in anti-parvalbumin at 1:1000. After rinsing in PBS, incubation in biotinylated anti-mouse IgG (1:500) was performed for 2 h. Sections were then rinsed in PBS and incubated in ABC complex (1:250) for 1 h. This step was followed by washing in PBS and then in Tris-HCl which preceded incubation in 0.05% DAB containing 1% H_2O_2 (1:100) for visualization of the reaction.

All the incubations were performed on a shaker at room temperature unless mentioned otherwise. Immunocytochemical reactions were accompanied by controls for specificity of the methods used. Sections were mounted on slides, dried, and coverslipped with entellan (Merck).

2.9
Morphometric Studies and Statistics

All sections were studied with the light microscope Olympus BX40 connected to image analyzer CUE-2. Stained neurons with a visible nucleus were marked manually with a black point in the studied plane. Points were summed and the density of cells was determined. ANOVA and Student's t test were applied for comparison of animals by sex, age, and experiment. Differences of $P<0.01$ were considered statistically significant.

3 Results

3.1
Bed Nucleus of the Stria Terminalis

3.1.1
Volume and Neuronal Density

On Nissl-stained preparations, we localized the BST and its main portions – medial, lateral, and ventral. Morphometric studies were done on the volume of the BST and neuronal densities in its subdivisions to compare both sexes and effect of androgens.

In coronary sections the BST (Fig. 3) was followed as the immediate continuation of the nucleus accumbens, which surrounds the anterior commissure. Caudally, the borders of the nucleus widened, while laterally it was confined by the internal capsule and striatum. Medially, the BST spanned to the septal nuclei and more caudally to the fornix and stria medullaris. The upper pole of the BST reached the inferior horn of the lateral ventricle. Ventral borders were not well delineated and gradually traversed into the innominate substance and into the preoptic area. Caudal levels of the BST ex-hausted among the caudal fibers of the stria terminalis and the rostral pole of the thalamus. The medial BST was situated dorsomedially. Mainly oval cell bodies with a small or medium size were present here. Some of the neurons were parallel with their long axis to the fibers of stria terminalis. In the posterior part a cluster of neurons could be discerned. Using the image analysis system Olympus CUE-2, the mean length of the long axis of neurons in the medial BST was determined to be 16.1±3.94 µm. The lateral BST was in close proximity to the internal capsule. It was transversed by the anterior commissure, which divided it into a dorsal and ventral part. Ovoid perikarya of small and medium size were found in the lateral BST. The long axis of neurons in the dorsal part measured 18.3±1.92 µm. Near the internal capsule, a cluster of smaller round neurons was found (14.6±1.14 µm). The ventral BST was localized immediately under the anterior commissure. It included predominantly small neurons with an oval or a fusiform shape. Some of them were orientated parallel to the fibers of the anterior commissure. Morphometric studies revealed that the mean size of the long axis of neurons in the ventral BST was 15.2±2.73 µm.

With the aid of the image analyzer, the borders of the BST were delineated in serial sections and the volume of the nucleus was determined in male and female rats (Table 3). No sex difference was found in either of the studied age groups – 6 days, 20 days, 3 months, and 1 year. Castrations of newborn and pubertal males, as well as

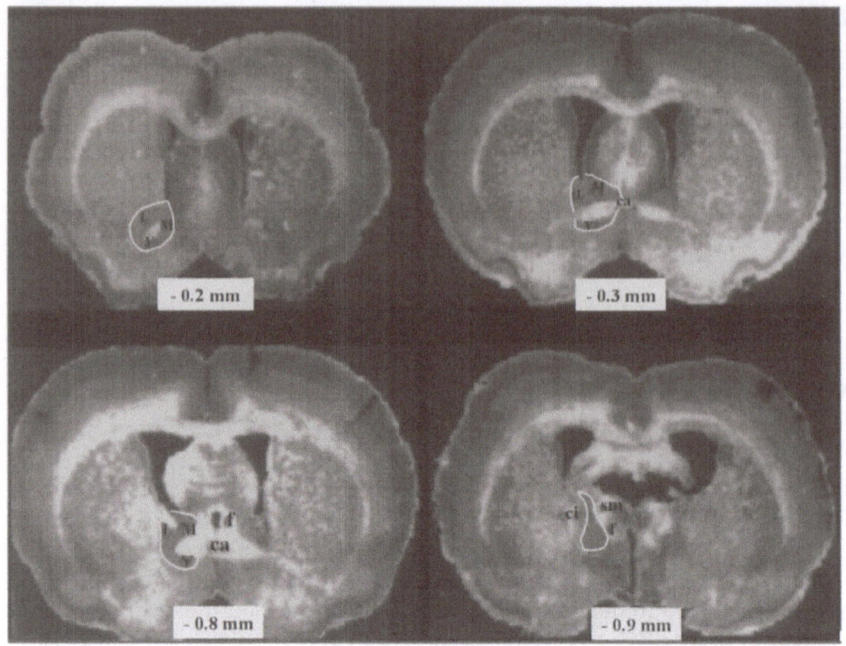

Fig. 3. Localization of the rat BST in coronary sections. Distance from bregma in mm. *L*, lateral BST; *M*, medial BST; *V*, ventral BST; *ca*, commissura anterior; *st*, stria terminalis; *ci*, capsula interna; *sm*, stria medullaris; *f*, fornix

treatment with aromatase inhibitor or estrogen antagonist during the first ten postnatal days, had no effect on BST volume (Table 4).

The density of neurons in the three main subdivisions of the BST in male and female rats was determined. Sex differences were not observed in either of the age groups examined (Table 5). The neuronal density in the BST gradually decreased with ageing (Fig. 4).

Experimental manipulations of hormonal environment in newborn male rats caused a significant decrease of the neuronal density only in the medial BST, while castrations in puberty had no significant effect (Fig. 5).

3.1.2
GABA-Immunoreactive Neurons

The BST is comparatively rich in GABA-immunoreactive neurons in all its subdivisions. They are usually small or medium sized. Their long axis is about 13.4±3.46 μm (mean±SD). The shape of GABA-immunoreactive perikarya in the BST is oval to multipolar (Fig. 6). In the rostral part of the medial BST, neurons are dispersed among the fibers of the stria terminalis. Caudally, a compact group of immunoreactive cells is observed that corresponds to the so-called encapsulated part of the medial subdivision. GABA-immunoreactive neurons in the lateral BST were equally dense through-

Table 3. Volumes of the BST of male (*M*) and female (*F*) rats in four age groups – 6 days (6), 20 days (20), 3 months (3), and 1 year (1) old. Data are mean±SD. In all cases there is no significant sex difference ($P>0.01$)

Volume of BST in mm³	M6	F6	M20	F20	M3	F3	M1	F1
	0.8344±0.058	0.7781±0.031	0.8844±0.071	0.8581±0.077	1.1257±0.079	1.0393±0.096	1.1765±0.11	1.07±0.07

Table 4. Volumes of the BST in 3-month-old intact male (*M3*) and female (*F3*) rats, as well as castrated newborn males (*CN*), castrated at puberty males (*CP*), treated with estrogen antagonist (*EA*) and aromatase inhibitor (*AI*) male rats. Data are mean±SD. In all cases there is no significant sex difference ($P>0.01$)

Volume of BST in mm³	M3	F3	CN	CP	EA	AI
	1.1257±0.079	1.0393±0.096	1.0745±0.064	1.1932±0.067	1.0996±0.028	1.1239±0.027

Table 5. Density of neurons per mm3 in the lateral (*BST-L*), medial (*BST-M*), and ventral (*BST-V*) bed nucleus of the stria terminalis. Data are mean±SD. In all cases there is no significant sex difference ($P>0.01$)

Age	BST-L		BST-M		BST-V	
	Male	Female	Male	Female	Male	Female
6 days	85125±4625	82750±505	73500±5500	62500±6750	84875±1625	80000±9000
20 days	75000±1250	73125±7578	57083±1573	56416±3502	76041±722	70083±3263
3 months	65833±1465	66416±1127	48250±1887	42416±1507	51916±1876	61083±6903
1 year	39250±7155	54083±7609	33833±5965	46333±6370	42500±5414	47333±4474

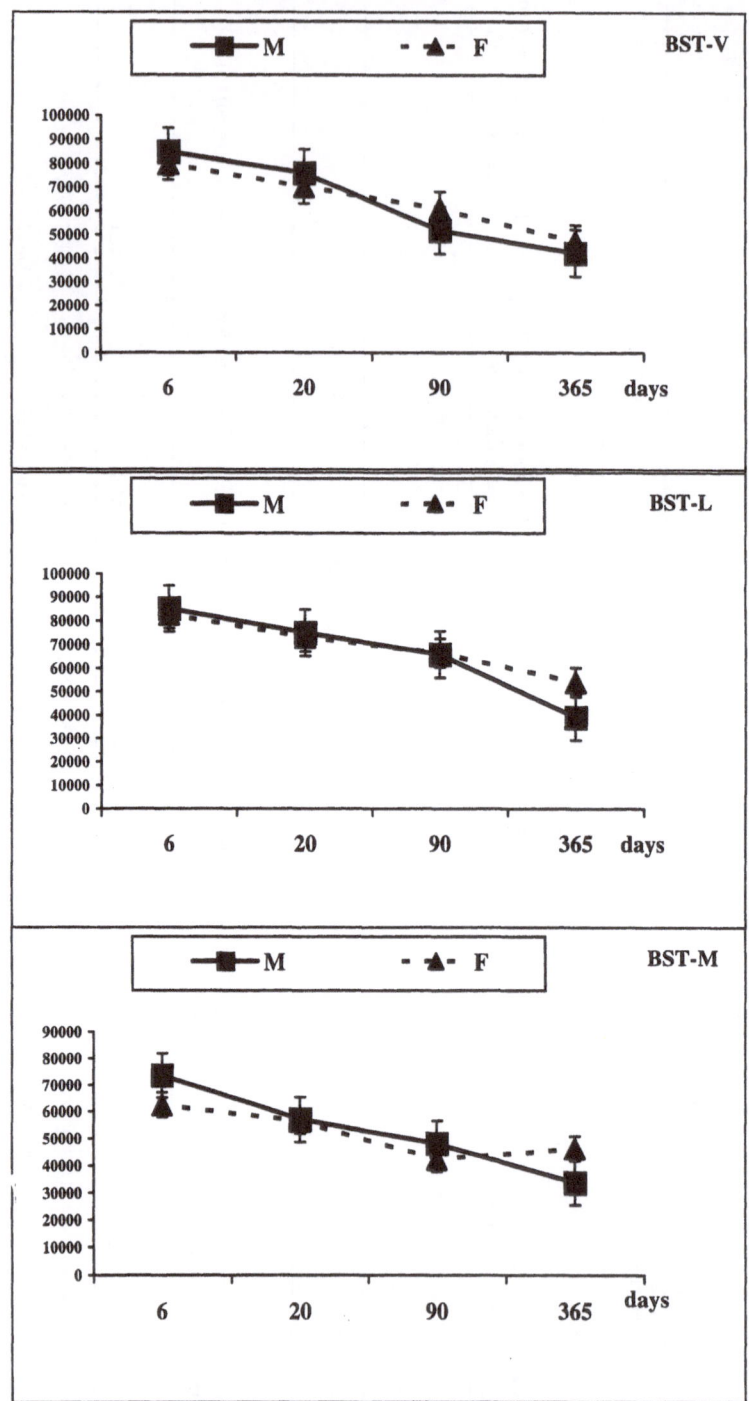

Fig. 4. Age-related changes in the neuronal density of the ventral (*BST-V*), lateral (*BST-L*), and medial (*BST-M*) bed nucleus of the stria terminalis of male and female rats. Data are mean±SEM

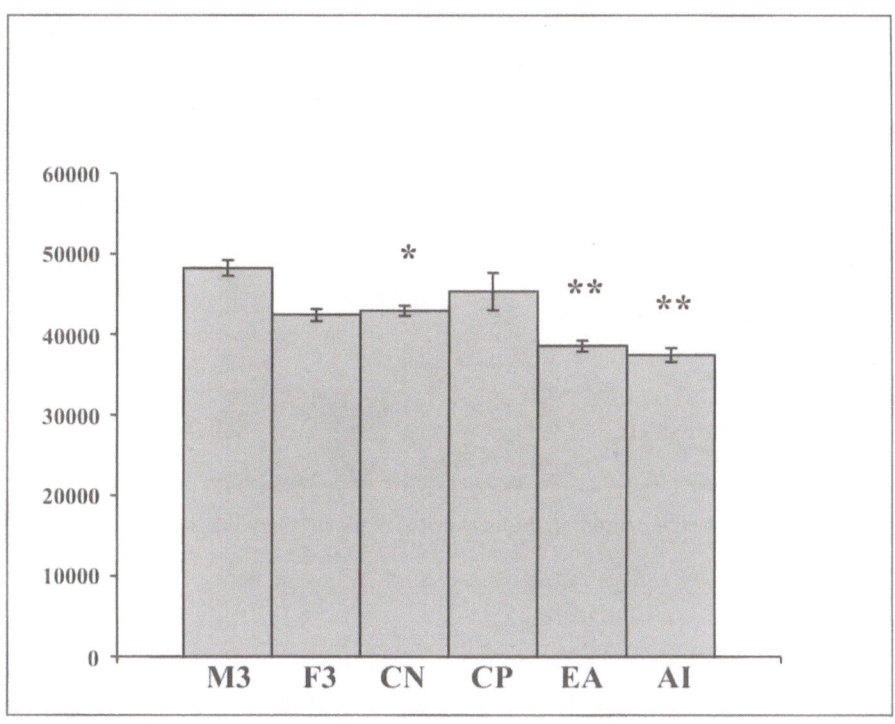

Fig. 5. Changes in the neuronal density of the medial BST after manipulations of the hormonal environment. Data are mean±SEM. *$P<0.01$ or **$P<0.001$ compared to M3. *M3*, intact male, 3 months old; *F3*, female, 3 months old; *CN*, 3-month-old male, castrated on first day of life; *CP*, 3-month-old male, castrated at puberty; *EA*, 3-month-old male, treated with estrogen antagonist in the first 10 days of life; *AI*, 3-month-old male, treated with aromatase inhibitor in the first 10 days of life

out its continuation. These were predominantly medium-sized perikarya. In the ventral BST, typical fusiform GABA-immunoreactive perikarya were observed. They were usually parallel to the fibers of the anterior commissure.

At 6 days of age no sex difference was established in the density of GABA-immunoreactive neurons in the subdivisions of the BST (Table 6). Sex difference was first observed in the medial BST at 20 days of age (Table 6). In postpubertal rats, sexual dimorphism of GABA-immunoreactive neurons was found in the three main subdivisions of the BST (Table 6). Female rats had a greater neuronal density than males.

Interestingly, the density of GABA-immunoreactive neurons in the subdivisions of the BST increased with ageing (Fig. 7). This tendency was observed in absolute numbers as well as in the percentage of GABA-immunoreactive cells (Table 7).

Changes in the density of GABA-immunoreactive neurons were examined after manipulations of the hormonal environment during the critical period and in puberty. Castration of newborn male rats caused an increase in the density of GABA-immunoreactive neurons at 3 months of age compared to intact males. It reached female levels (Fig. 8). Castration of males in puberty also caused an increase of GABA-im-

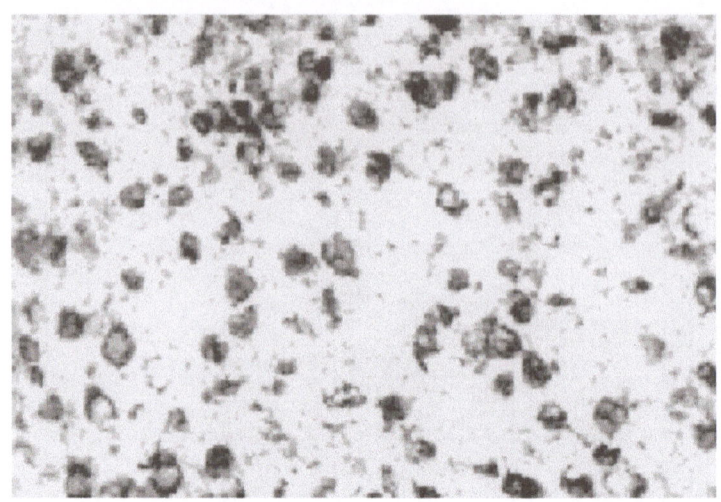

Fig. 6. GABA-immunoreactive neurons in the bed nucleus of the stria terminalis. ×880

Table 6. Density of GABA-immunoreactive neurons in the lateral (*BST-L*), medial (*BST-M*), and ventral (*BST-V*) bed nucleus of the stria terminalis of male and female rats at 6 days (6), 20 days (20), 3 months (3), and 1 year (1) of age. Data are mean±SD

GABA	BST-L-6	BST-M-6	BST-V-6
Male	17666±190	17750±4250	17083±3820
Female	18333±190	17187±1562	17666±401
	BST-L-20	**BST-M-20**	**BST-V-20**
Male	21771±1630	14541±288	16896±1941
Female	24500±2741	19354±781**	19062±4173
	BST-L-3	**BST-M-3**	**BST-V-3**
Male	25000±1082	16875±533	23229±566
Female	29333±803*	21937±676*	29166±1127*
	BST-L-1	**BST-M-1**	**BST-V-1**
Male	31833±1464	22833±1343	37125±780
Female	45750±661**	37125±780**	42458±505*

*Sex difference at $P<0.01$; **sex difference at $P<0.001$.

Table 7. Age-related changes in the percentage of GABA-immunoreactive neurons in the medial (*BST-M*), lateral (*BST-L*), and ventral (*BST-V*) bed nucleus of the stria terminalis from the total neuronal population as measured in Nissl-stained preparations

% of GABA-immunoreactive neurons	BST-L		BST-M		BST-V	
	Male	Female	Male	Female	Male	Female
6 days	20.75	22.15	24.14	27.50	20.13	22.08
20 days	29.03	33.50	25.47	34.30	22.22	27.20
3 months	37.97	44.17	34.97	51.72	44.74	47.75
1 year	81.10	84.59	67.49	80.13	87.35	89.7

munoreactive neuronal density in the three main subdivisions of the BST (Fig. 8). Estrogen antagonist or aromatase inhibitor treatment caused a significant increase of GABA-immunoreactive neuronal density in the medial BST, but not in the ventral and lateral BST (Fig. 8).

3.1.3
Leucine-Enkephalin-Immunoreactive Neurons

Leucine-enkephalin-immunoreactive neurons in the BST are predominantly small or medium sized (long axis 13.3±2.34 µm). They have multipolar to oval perikarya (Fig. 9). Fusiform cell bodies could be observed mainly in the ventral subdivision of the BST, usually parallel to the anterior commissure, as well as in the medial BST, parallel to the fibers of the stria terminalis.

Sexual dimorphism in the density of leucine-enkephalin-immunoreactive neurons was found in the lateral BST. Males had a greater density of leucine-enkephalin-immunoreactive neurons in all the studied age groups compared to females (Table 8). Sex differences were not found in the medial and ventral BST (Table 8).

The density of leucine-enkephalin-immunoreactive nerve cells decreased with ageing, corresponding to the decrease of the total neuronal density in the BST as measured in Nissl-stained sections (Fig. 10).

Experimental castrations of newborn male rats caused a decrease in the density of leucine-enkephalin-immunoreactive neurons in the lateral BST, yet not reaching female levels (Fig. 11). Castration of pubertal male rats did not affect the density of leucine-enkephalin-immunoreactive neurons in the subdivisions of the BST (Fig. 11). Treatment with estrogen antagonist or aromatase inhibitor of newborn males during the critical period of sexual differentiation caused a significant decrease in the density of leucine-enkephalin-immunoreactive neurons in the lateral BST compared to untreated males at 3 months of age. In the same period they did not reach female levels (Fig. 11). In the medial and ventral BST, castrations and treatment with estrogen

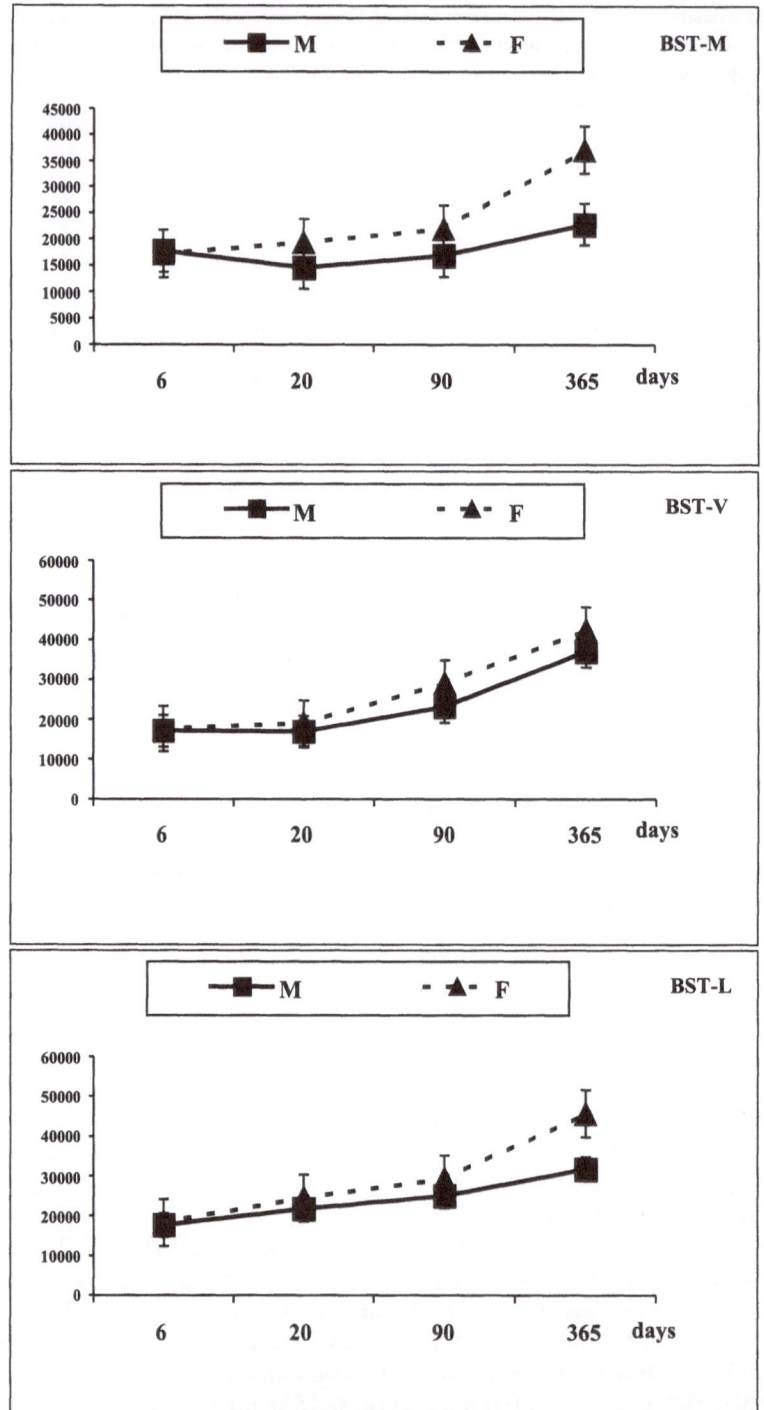

Fig. 7. Age-related changes in the density of GABA-immunoreactive neurons in the ventral (*BST-V*), lateral (*BST-L*), and medial (*BST-M*) bed nucleus of the stria terminalis of male and female rats. Data are mean±SEM

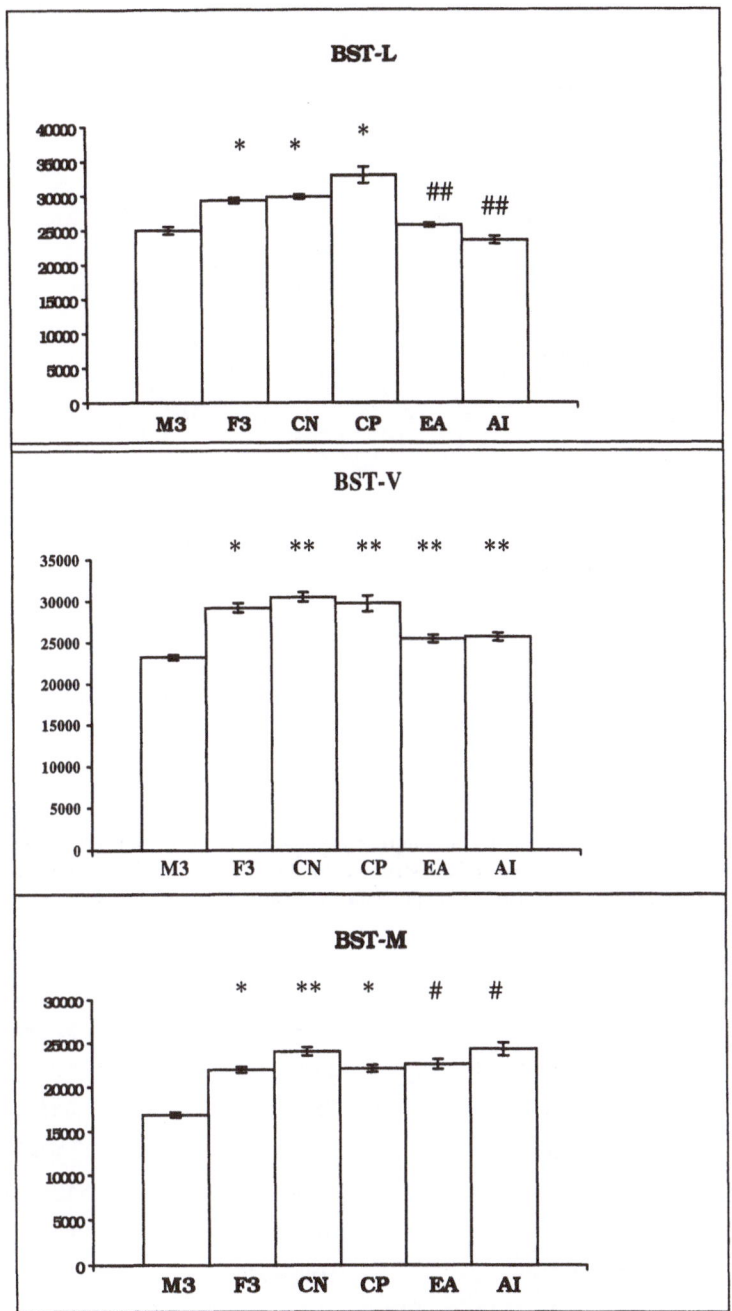

Fig. 8. Effect of manipulations of hormonal environment on the density of GABA-immunoreactive neurons in the medial (*BST-M*), lateral (*BST-L*), and ventral (*BST-V*) bed nucleus of the stria terminalis. Data are mean±SEM. *$P<0.01$ or **$P<0.001$ compared to M3; #$P<0.01$ or ##$P<0.001$ compared to F3. *M3*, intact male, 3 months old; *F3*, female, 3 months old; *CN*, 3-month-old male, castrated on first day of life; *CP*, 3-month-old male, castrated at puberty; *EA*, 3-month-old male, treated with estrogen antagonist in the first 10 da

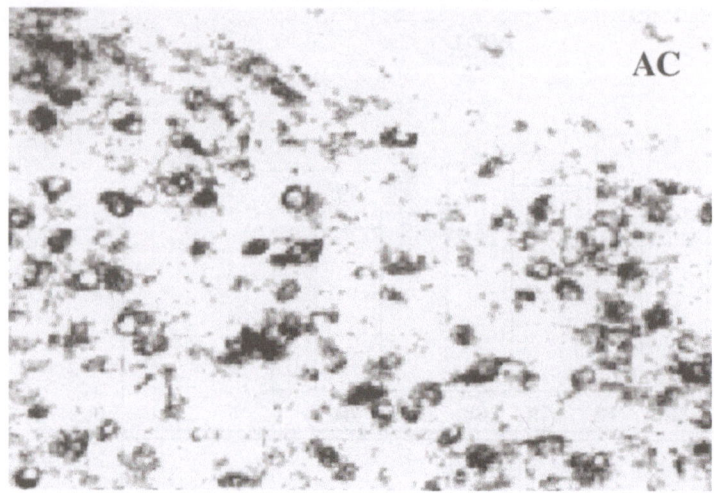

Fig. 9. Leucine-enkephalin-immunoreactive neurons in the bed nucleus of the stria terminalis of the rat. *AC*, anterior commissure. ×880

Table 8. Density of leucine-enkephalin-immunoreactive neurons in the lateral (*BST-L*), medial (*BST-M*), and ventral (*BST-V*) bed nucleus of the stria terminalis of male and female rats at 6 days (6), 20 days (20), 3 months (3), and 1 year (1) of age. Data are mean±SD

Leucine-enkephalin	BST-L-6	BST-M-6	BST-V-6
Male	29000±250	17683±633	32866±1220
Female	26916±629*	16888±315	33083±2020
	BST-L-20	**BST-M-20**	**BST-V-20**
Male	30250±500	17950±327	33400±687
Female	17641±534**	17033±800	32700±638
	BST-L-3	**BST-M-3**	**BST-V-3**
Male	26000±250	16000±250	33333±946
Female	16041±616*	12500±572	28500±1500
	BST-L-1	**BST-M-1**	**BST-V-1**
Male	17083±190	12604±1644	23021±996
Female	10125±495**	8479±632	20541±1018

*Sex difference at $P<0.01$; **sex difference at $P<0.001$.

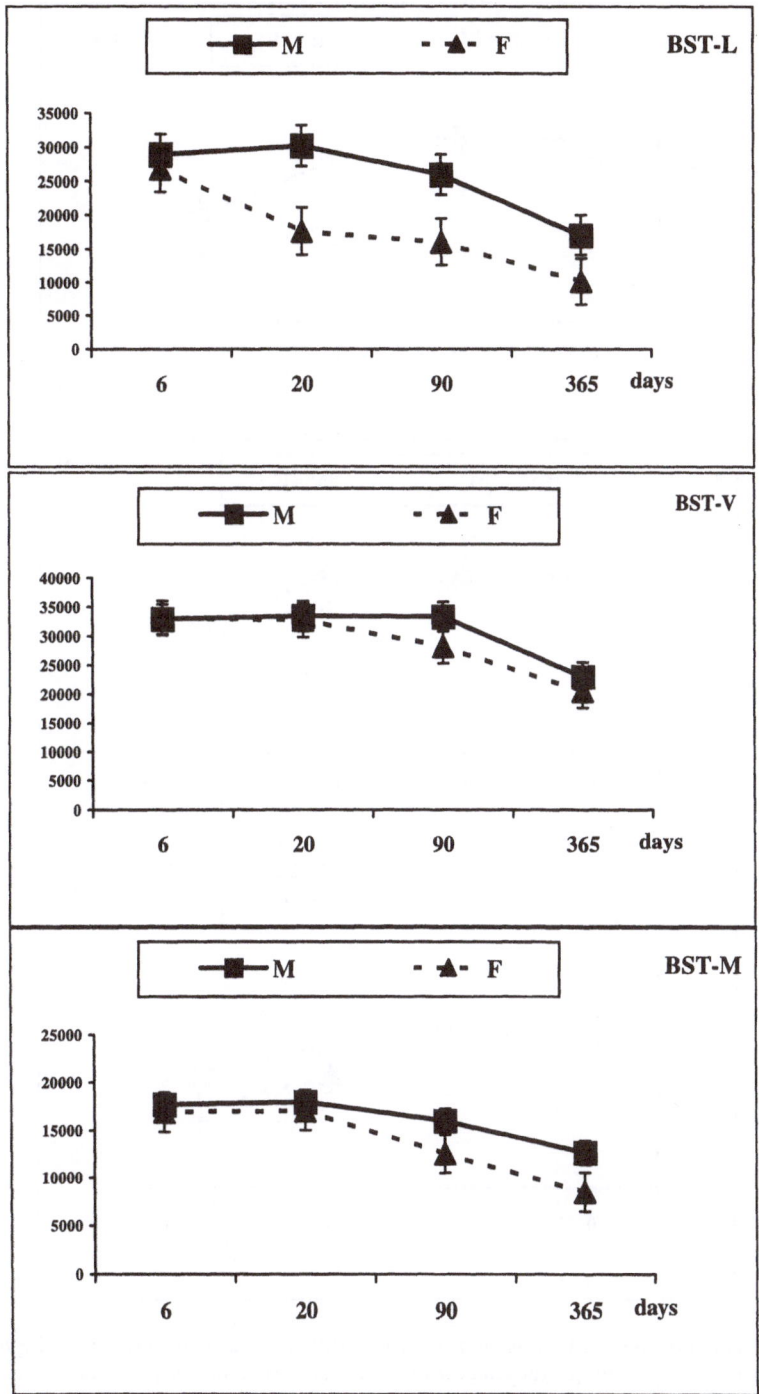

Fig. 10. Age-related decrease in the density of leucine-enkephalin-immunoreactive neurons in the ventral (*BST-V*), lateral (*BST-L*), and medial (*BST-M*) bed nucleus of the stria terminalis of male and female rats. Data are mean±SEM

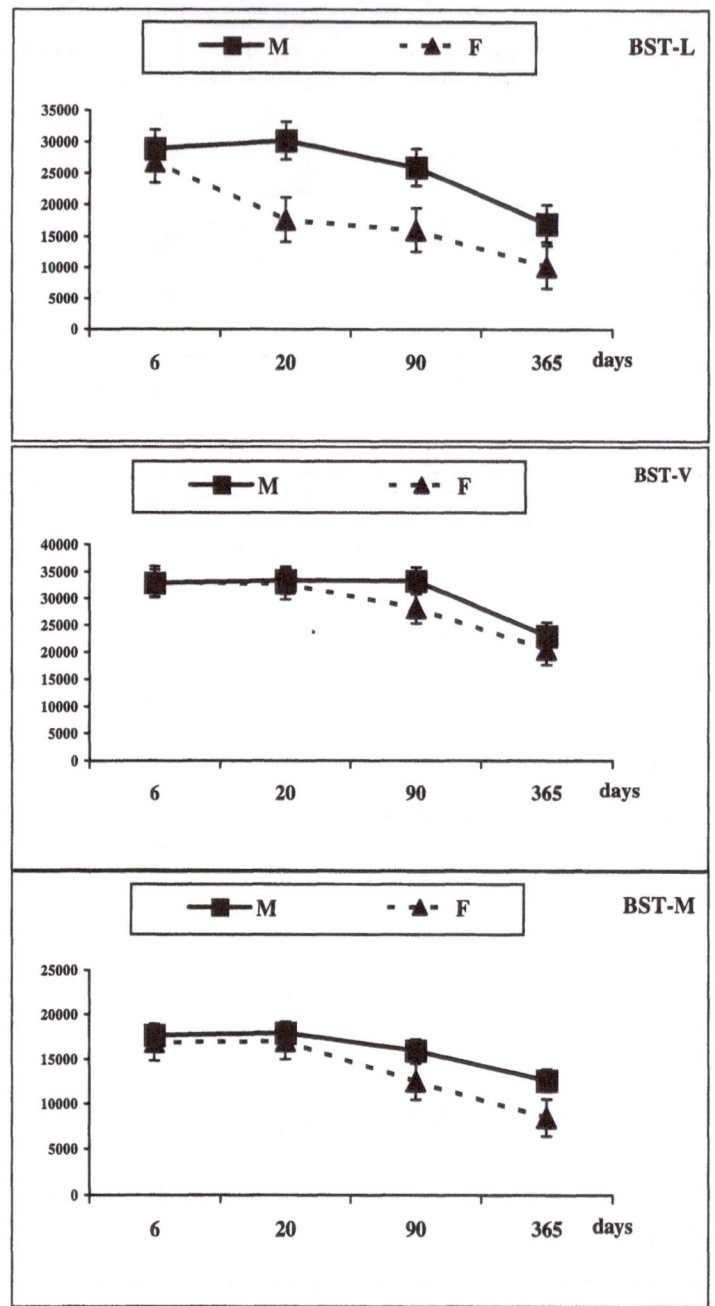

Fig. 11. Effect of manipulations of hormonal environment on the density of leucine-enkephalin-im-munoreactive neurons in the lateral bed nucleus of the stria terminalis (*BST-L*). Data are mean±SEM. *$P<0.01$ or **$P<0.001$ compared to M3; # $P<0.01$ or ##$P<0.001$ compared to F3. *M3*, intact male, 3 months old; *F3* female, 3 months old; *CN*, 3-month-old male, castrated on first day of life; *CP*, 3-month-old male, castrated at puberty; *EI*, 3-month-old male, treated with estrogen antagonist in the first 10 days of life; *AI*, 3-month-old male, treated with aromatase inhibitor in the first 10 days of life

Table 9. Density of leucine-enkephalin-immunoreactive neurons in the medial (*BST-M*) and ventral (*BST-V*) bed nucleus of the stria terminalis. Data are mean±SD

Leucine-enkephalin	M3	F3	CN	CP	EA	AI
BST-M	16000±250	12500±572	14156±1017	14812±1477	12543±1512	16593±2005
BST-V	33333±946	28250±1500	27812±2375	29875±2989	29687±6111	26488±2809

M3, intact male, 3 months old; *F3*, female, 3 months old; *CN*, 3-month-old male, castrated on first day of life; *CP*, 3-month-old male, castrated at puberty; *EA*, 3-month-old male, treated with estrogen antagonist in the first 10 days of life; *AI*, 3-month-old male, treated with aromatase inhibitor in the first 10 days of life.

antagonist or aromatase inhibitor had no effect on the density of leucine-enkephalin-immunoreactive neurons (Table 9).

3.1.4
Parvalbumin-Immunoreactive Neurons

No parvalbumin-immunoreactive neurons were observed in the BST of 6-day-old male and female rats. The expression of parvalbumin-positive perikarya in 20-day-old rats was scarce in the BST. Parvalbumin-immunoreactive cell bodies were observed in the ventrolateral nucleus of the BST according to the description of Paxinos and Watson (1986). These were small immunopositive perikarya with a long axis

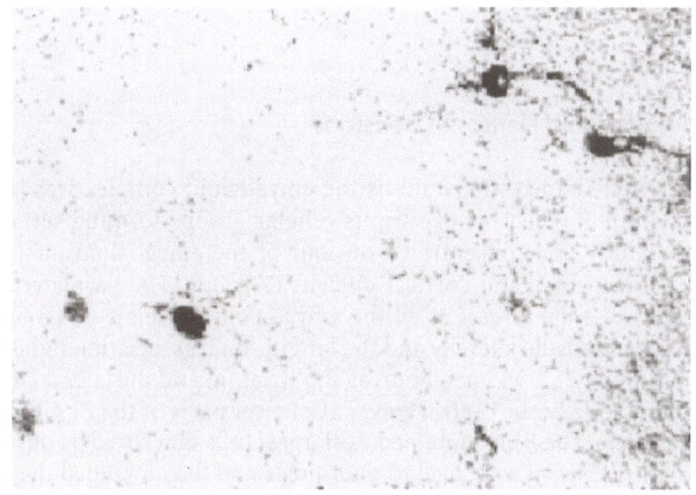

Fig. 12. Parvalbumin-immunoreactive neurons in the bed nucleus of the stria terminalis. ×880

Table 10. Density of parvalbumin-immunoreactive neurons in the ventrolateral part of the bed nucleus of the stria terminalis. Data are mean±SD

Parvalbumin	6 days	20 days	3 months	1 year	CN	CP	EA	AI
Male	0	4717±104	4225±108	2200±100	4425±108	3900±100	4108±101	3850±150
Female	0	4425±108	3817±202	1517±292				

CN, 3-month-old male, castrated on first day of life; CP, 3-month-old male, castrated at puberty; EA, 3-month-old male, treated with estrogen antagonist in the first 10 days of life; AI, 3-month-old male, treated with aromatase inhibitor in the first 10 days of life.

(11.5±1.52 µm) that were not intensely marked and did not show their dendrites well (Fig. 12). The morphometric study of the density of parvalbumin-immunoreactive neurons in the region of their highest concentration in the BST did not establish any sex difference in the age groups studied (Table 10). We did not find any effect of androgen levels on the density of parvalbumin-immunoreactive neurons in castrated newborn and pubertal male rats compared to intact males at the age of 3 months. Treatment with estrogen antagonist or aromatase inhibitor of newborn males had no effect on parvalbumin-positive neurons either (Table 10).

As we followed the changes in the density of parvalbumin-immunoreactive neurons with ageing, it was established that parvalbumin positivity was settled between 6 and 20 days of age in both sexes. In the next postpubertal ages – 3 months and 1 year – the density of parvalbumin-immunoreactive neurons gradually decreased (Table 10).

3.2
The Amygdala

3.2.1
Volumes and Neuronal Densities

On Nissl-stained preparations the amygdaloid complex was localized ventral to the striatum. It consisted of various cellular groups forming several amygdaloid nuclei. Our study was concentrated on four of the amygdaloid nuclei, i.e., the basolateral, central, medial, and cortical nucleus (Fig. 13). The basolateral amygdaloid nucleus (BLA) was the largest of all the amygdaloid nuclei. It was found surrounded by the external capsule laterally and the longitudinal association bundle medially. The upper pole of the BLA was neighboring the striatum and the lateral amygdaloid nucleus, and more caudally, the central amygdala. Lower parts of the BLA bordered the basomedial amygdala. The BLA contained the largest cells that could be observed in the amygdala. Their perikarya were oval to multipolar and they included the so-called magnocellular part of the nucleus. The long axis of these neurons measured 19.5±2.91 µm.

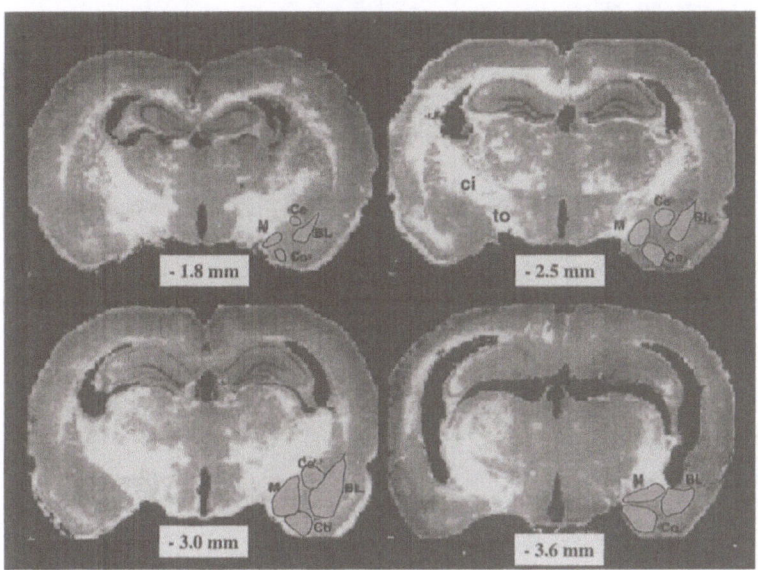

Fig. 13. Localization of the rat amygdala in coronary sections. Distance from bregma in mm. *M*, medial amygdala; *Co*, cortical amygdala; *Ce*, central amygdala; *BL*, basolateral amygdala; *ci*, capsula interna; *to*, tractus opticus

Together with them, smaller round to oval neurons with a long axis of 15.3±3.04 μm were observed in the parvicellular part of the BLA. The central nucleus (CeA) was surrounded by the fibers of the longitudinal association bundle laterally and ventrally, and by the stria terminalis medially. The CeA was situated ventrally from the striatum and laterally from the medial amygdaloid nucleus (MeA). A gradual transition between the anterior amygdaloid area and the CeA was found rostrally and between the CeA and the putamen caudally. Neurons in the CeA were predominantly small or medium sized. Smaller cells could be observed in the lateral part of the nucleus (long axis 14.5±1.35 μm), while in the medial part medium-sized neurons dominated (19.3±2.6 μm). However, the transition between the two parts of the nucleus was not very well demarcated. The MeA was observed laterally from the optic tract and ventrally to the CeA, the innominate substance, and the putamen. The lateral border of the MeA was next to the basomedial and cortical amygdala. Small oval or fusiform neurons were densely packed within the MeA. The long axis of the cellular perikarya measured 16.0±3.1 μm. The cortical amygdaloid nucleus (CoA) was situated most superficially of all the amygdaloid nuclei, and laterally it faded into the piriform cortex. The CoA did not show a well-preserved layer structure and at caudal levels it formed an ovoid mass that could readily be distinguished from the neighboring periamygdaloid and piriform cortex. Neuronal perikarya in the CoA were predominantly small and oval with a long axis of 15.9±2.94 μm.

With the aid of the image analyzer Olympus CUE-2, we delineated the borders of each of the four studied amygdaloid nuclei. Thus the area of the nucleus in each serial section was determined and the volume of the nucleus was calculated. Male and

Table 11. Volume of the cortical (*CoA*), central (*CeA*), basolateral (*BLA*), and medial (*MeA*) amygdala of intact and hormonally manipulated male and female rats. Data are mean±SD

Volume in mm³	CoA		CeA		BLA		MeA	
	Male	Female	Male	Female	Male	Female	Male	Female
6 days	0.7226±0.03	0.5781±0.06	0.7444±0.04	1.1052±0.07	0.9261±0.02	0.6559±0.04	0.817±0.067	0.722±0.01
20 days	0.8942±0.01	0.6033±0.02	0.9256±0.08	1.5528±0.08	1.3021±0.07	1.0500±0.12	1.582±0.04	1.055±0.05**
3 months	0.9622±0.04	0.9276±0.03	1.1472±0.10	1.5063±0.09	1.2833±0.04	0.8888±0.06	1.555±0.07	1.042±0.06**
1 year	0.8506±0.03	0.7401±0.04	1.0948±0.06	1.3807±0.07	1.1202±0.15	0.8852±0.11	1.52±0.1	0.966±0.03*
CN	0.7616±0.06		1.1729±0.03		1.3565±0.06		1.165±0.11*	
CP	0.8929±0.01		1.0729±0.05		1.5794±0.26		1.158±0.16	
EA	0.8768±0.03		1.0629±0.05		1.3777±0.20		1.104±0.01**	
AI	0.8721±0.02		0.9344±0.08		1.4880±0.30		1.081±0.02**	

CN, 3-month-old male, castrated on first day of life; *CP*, 3-month-old male, castrated at puberty; *EA*, 3-month-old male, treated with estrogen antagonist in the first 10 days of life; *AI*, 3-month-old male, treated with aromatase inhibitor in the first 10 days of life.
*P<0.01 or **P<0.001 compared to intact male of the same age.

Table 12. Density of neurons per mm3 in the cortical (*CoA*), central (*CeA*), basolateral (*BLA*), and medial (*MeA*) amygdala of male and female rats at 6 days, 20 days, 3 months, and 1 year of age. Data are mean±SD

Neuronal density	CoA		CeA		BLA		MeA	
	Male	Female	Male	Female	Male	Female	Male	Female
6 days	62500±1500	52500±6750	75875±3125	69750±2500	66000±2500	62500±3500	83125±7625	73375±2625
20 days	52750±1000	41916±381**	58125±2831	51250±2704	42250±3750	36833±629	63250±2750	49125±1125*
3 months	53187±3071	43375±2066*	46812±2749	42750±889	44000±1903	41875±1198	56062±2695	44812±2066**
1 year	39500±901	33333±1010*	37083±1607	31916±1909	28500±2610	25750±1750	47916±1010	42000±250**

*Sex difference at *P*<0.01; **sex difference at *P*<0.001.

Table 13. Density of neurons in the central (*CeA*) and basolateral (*BLA*) amygdala of castrated newborn (*CN*) males, castrated males at puberty (*CP*), males treated with estrogen antagonist in the first 10 days of life (*EA*), and males treated with aromatase inhibitor in the first 10 days of life (AI). Data are mean±SD

Neuronal density	CeA	BLA
CN	44312±8226	40687±2697
CP	41875±1506	42875±4939
EA	40375±5222	42581±2711
AI	43500±1968	40000±1409

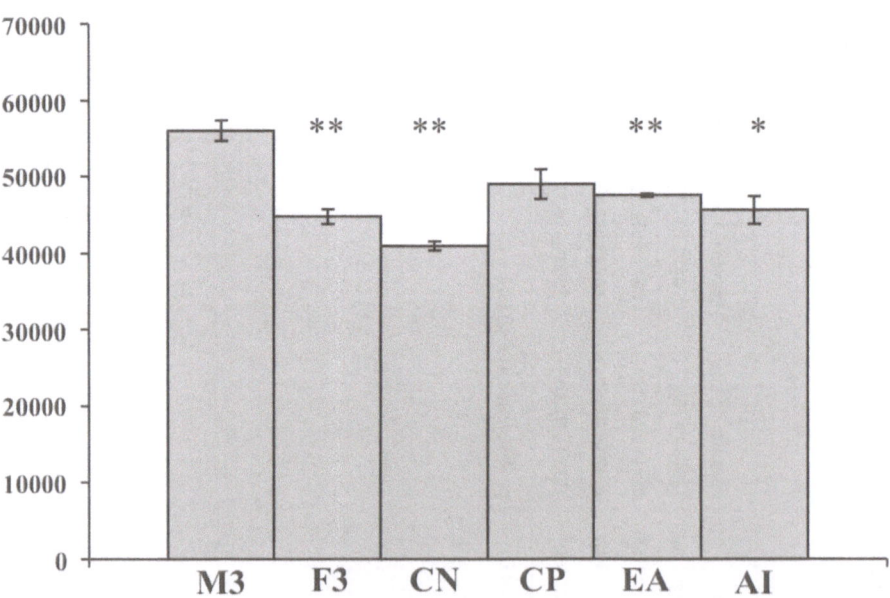

Fig. 14. Effect of manipulations of hormonal environment on the density of neurons in the medial amygdala. Data are mean±SEM. *$P<0.01$ or **$P<0.001$ compared to M3. *M3*, intact male, 3 months old; *F3*, female, 3 months old; *CN*, 3-month-old male, castrated on first day of life; *CP*, 3-month-old male, castrated at puberty; *EA*, 3-month-old male, treated with estrogen antagonist in the first 10 days of life; *AI*, 3-month-old male, treated with aromatase inhibitor in the first 10 days of life

CoA

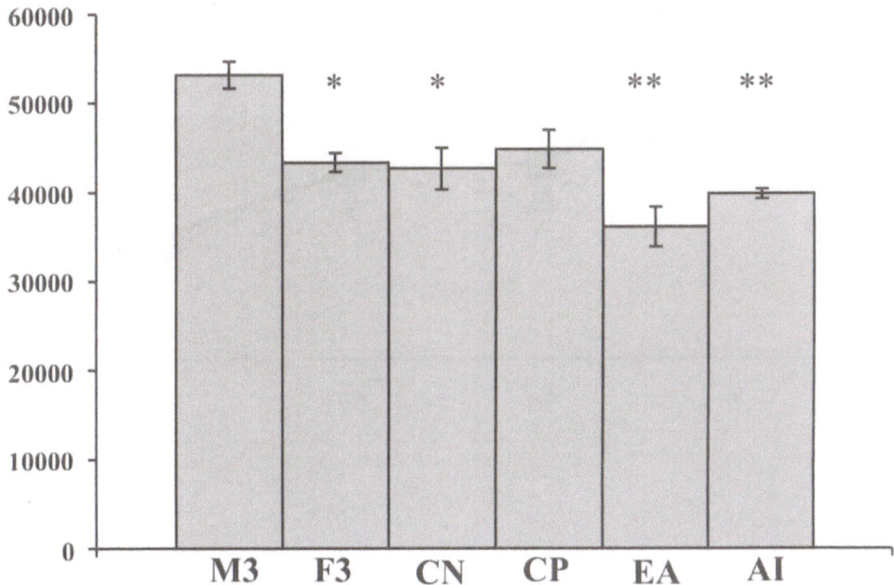

Fig. 15. Effect of manipulations of hormonal environment on the density of neurons in the cortical amygdala. Data are mean±SEM. *P<0.01 or **P<0.001 compared to M3. *M3*, intact male, 3 months old; *F3*, female, 3 months old; *CN*, 3-month-old male, castrated on first day of life; *CP*, 3-month-old male, castrated at puberty; *EA*, 3-month-old male, treated with estrogen antagonist in the first 10 days of life; *AI*, 3-month-old male, treated with aromatase inhibitor in the first 10 days of life

female volumes in the four age groups were compared. At 6 days of age, no sex difference was found in the volume of either of the amygdaloid nuclei (Table 11). The volume of the MeA in 20-day-old male rats was greater than in females of the same age (Table 11). Volumes of the CoA, CeA, and BLA did not differ between sexes at 20 days of age and in the next postpubertal ages studied (Table 11). The sex difference in the volume of the MeA was preserved in 3-month- and 1-year-old rats (Table 11).

The experimental manipulations of the hormonal environment of male rats had no effect on the volume of the CoA, CeA, and BLA (Table 11). Castration of pubertal males caused a decrease of the MeA volume but it was not significantly different from that of intact males in the third month of life (Table 11). Changing the androgen influence during the critical period of sexual differentiation by castration of newborn males and treatment with the estrogen antagonist or the aromatase inhibitor had unidirectional effects. The volume of the MeA decreased significantly compared to intact males at the age of 3 months and reached female size (Table 11).

Morphometric studies of the neuronal density in the four amygdaloid nuclei studied revealed the existence of sex differences. It was represented differently in the

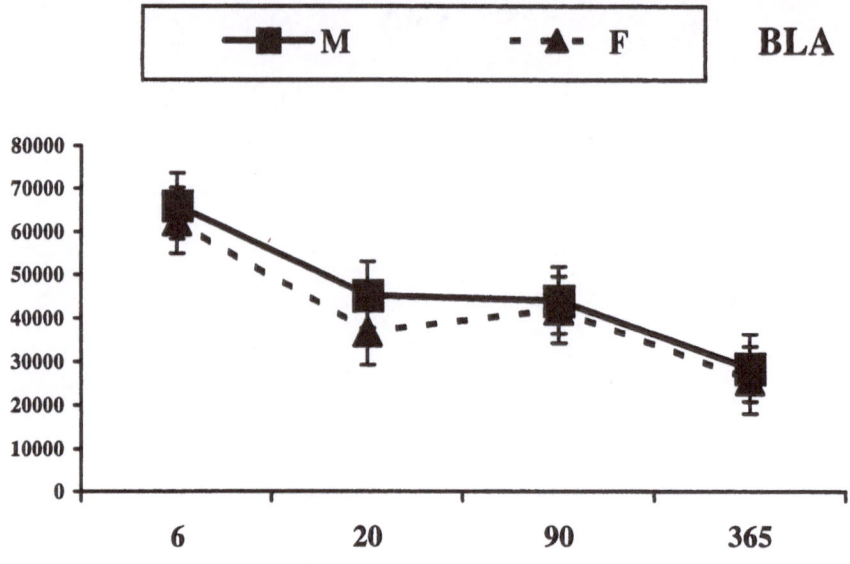

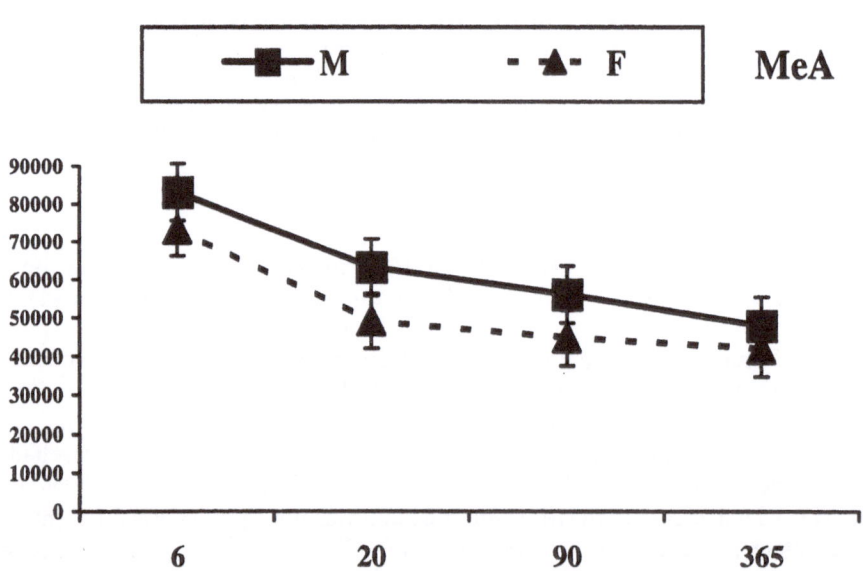

Fig. 16. Age-related changes in the neuronal density of male and female medial (*MeA*), cortical (*CoA*), central (*CeA*), and basolateral (*BLA*) amygdala. Data are mean±SEM

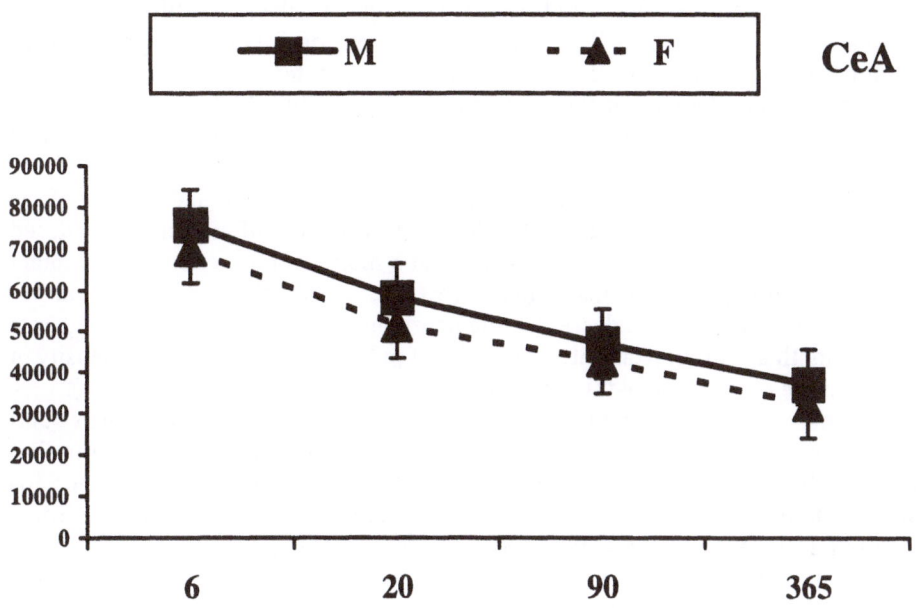

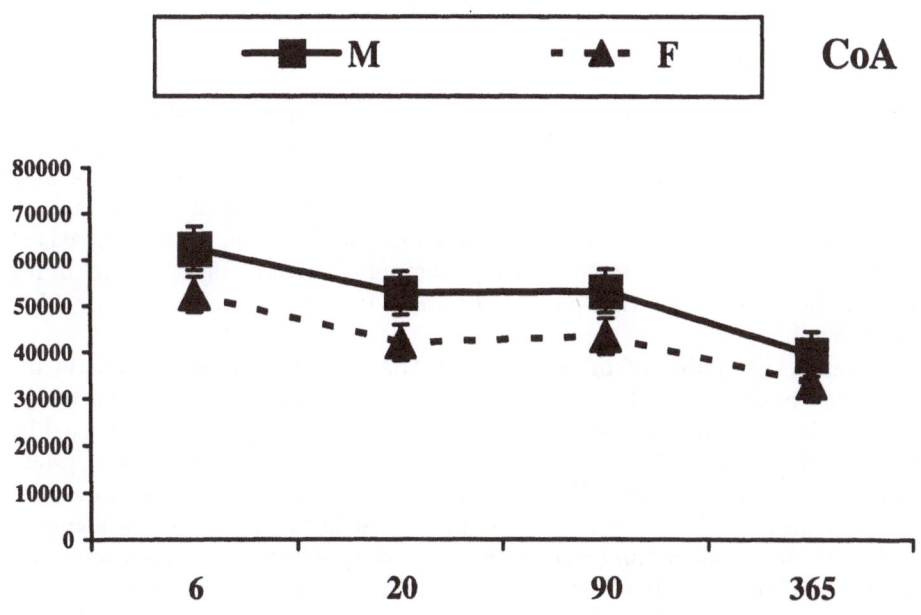

various parts of the amygdala and with different terms of expression. In the MeA and the CoA, sex differences were not observed at the age of 6 days. Sex difference in the neuronal density of these nuclei was found at 20 days of age and was preserved in 3-month- and 1-year-old rats. In all cases, males had greater neuronal density in the MeA and CoA than females of the corresponding age (Table 12). In the CeA and the BLA, no sex difference in the neuronal density was found in any of the ages studied (Table 12).

The experimental manipulations of the hormonal environment had diverse effects on the various amygdaloid nuclei. In the MeA, castration of newborn male rats caused a significant decrease in the neuronal density at 3 months of age compared to intact males of the same age. Analogously, injecting estrogen antagonist or aromatase inhibitor in the first 10 days of life decreased the neuronal density in the MeA of male rats in the 3-month age group. Castration of males at puberty did not affect the density of neurons in the MeA compared to intact males in the third month of life (Fig 14).

The same results were established in the CoA. Castration of newborn males and treatment with estrogen antagonist or aromatase inhibitor of male rats in the first 10 days of life caused a significant decrease in the neuronal density of the nucleus compared to intact males. Castration of pubertal males had no effect on the neuronal density in the CoA, and the sex difference was preserved (Fig. 15).

Hormonal manipulations had no significant effect on the neuronal density of the CeA and BLA. Neither castrations of newborns or pubertal males nor treatment with estrogen antagonist or aromatase inhibitor of newborn males caused changes in the density of neurons in these amygdaloid nuclei (Table 13).

Comparison of the age groups revealed a decrease of the neuronal density in the MeA, CoA, CeA, and BLA with ageing in both sexes (Fig. 16).

3.2.2
GABA-Immunoreactive Neurons

In all the amygdaloid nuclei, intense GABA-immunoreactivity was observed. In the MeA, CoA, and CeA, an intensely labeled neuropil and small or medium-sized neuronal perikarya were seen (Fig. 17). GABA-immunoreactive neurons in the BLA were mainly medium sized and single large ones, which were evenly dispersed throughout the nucleus (Fig. 18), were present. Morphometric measurement of the long axis of GABA-immunoreactive neuronal perikarya showed a length of 15.8±2.96 µm in the MeA; in the CoA, 16.0± 1.94 µm; in the CeA, 16.2±3.92 µm; and in the BLA, 19.2±3.48 µm.

Morphometric studies of the density of GABA-immunoreactive neurons in the amygdaloid nuclei revealed the presence of sex differences. These differences appeared in different time periods of individual life and were specifically affected by hormonal environment. The comparison of 6-day-old male and female rats established that there was no sex difference in the density of GABA-immunoreactive neurons in the MeA, CoA, CeA, and BLA (Table 14). At 20 days of age, sex difference in the density of GABA-immunoreactive nerve cells appeared in the MeA and CoA (Table 14). At 3 months and at 1 year of age, sex differences were detected in the four amygdaloid nuclei examined (Table 14). Female rats had greater densities of GABA-immunoreactive neurons than males of the same age.

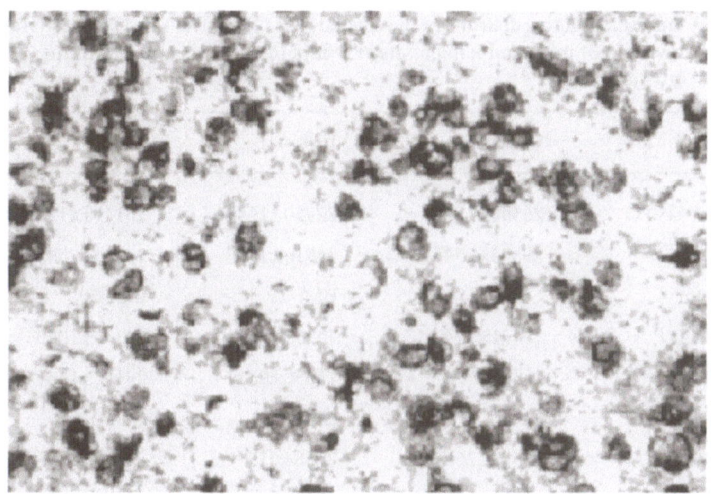

Fig. 17. GABA-immunoreactive neurons in the central amygdala. ×880

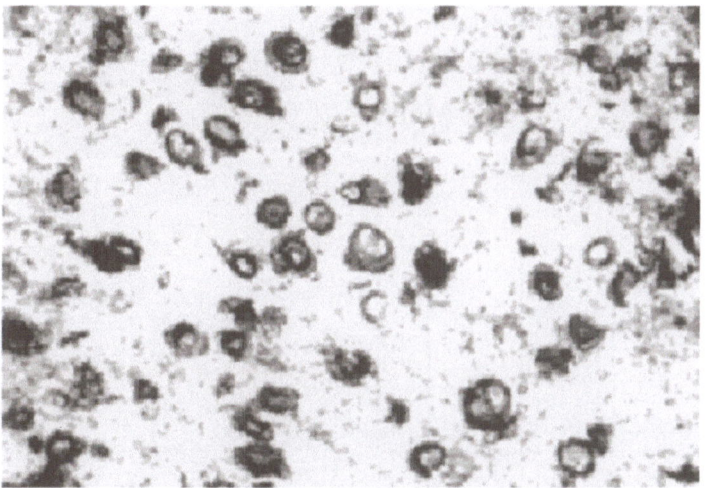

Fig. 18. GABA-immunoreactive neurons in the basolateral amygdala. ×880

Experimental manipulations of the hormonal environment had diverse effects on the density of GABA-immunoreactive neuronal populations in the different amygdaloid nuclei. In the MeA, castration of newborn male rats caused a significant increase in the density of GABA-immunoreactive neurons at 3 months of age compared to intact males of the same age. Yet, castrated males did not reach female densities, which was significantly different. Castration of male rats at puberty had the

Table 14. Density of GABA-immunoreactive neurons in the medial (*MeA*), cortical (*CoA*), basolateral (*BLA*), and central (*CeA*) amygdala of male and female rats at 6 days (6), 20 days (20), 3 months (3), and 1 year (1) of age. Data are mean±SD

GABA	MeA-6	CoA-6	BLA-6	CeA-6
Male	35500±6250	26500±1500	22950±1688	26141±619
Female	36916±1421	29583±5547	24916±1376	27083±2005
	MeA-20	CoA-20	BLA-20	CeA-20
Male	26003±442	18441±724	14133±561	27333±1040
Female	31250±901**	25550±576**	15658±932	28166±1664
	MeA-3	CoA-3	BLA-3	CeA-3
Male	18854±376	13583±260	7958±915	21729±1598
Female	34416±1181**	21500±1415**	11437±286*	29250±100*
	MeA-1	CoA-1	BLA-1	CeA-1
Male	9291±157	13333±438	6125±433	14812±1167
Female	16479±718*	22271±972*	8125±216*	24708±314*

*Sex difference at $P<0.01$; **sex difference at $P<0.001$.

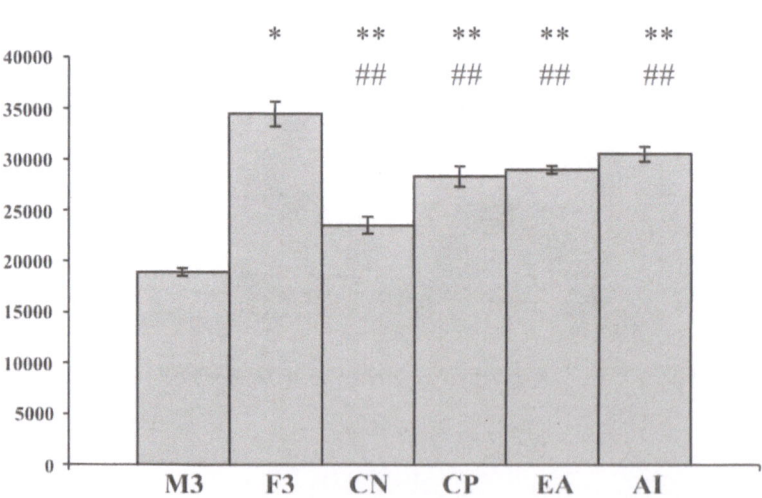

Fig. 19. Effect of manipulations of hormonal environment on the density of GABA-immunoreactive neurons in the medial amygdala (*MeA*). Data are mean±SEM. **$P<0.001$ compared to M3; ##$P<0.001$ compared to F3. *M3*, intact male, 3 months old; *F3*, female, 3 months old; *CN*, 3-month-old male, castrated on first day of life; *CP*, 3-month-old male, castrated at puberty; *EA*, 3-month-old male, treated with estrogen antagonist in the first 10 days of life; *AI*, 3-month-old male, treated with aromatase inhibitor in the first 10 days of life

CoA

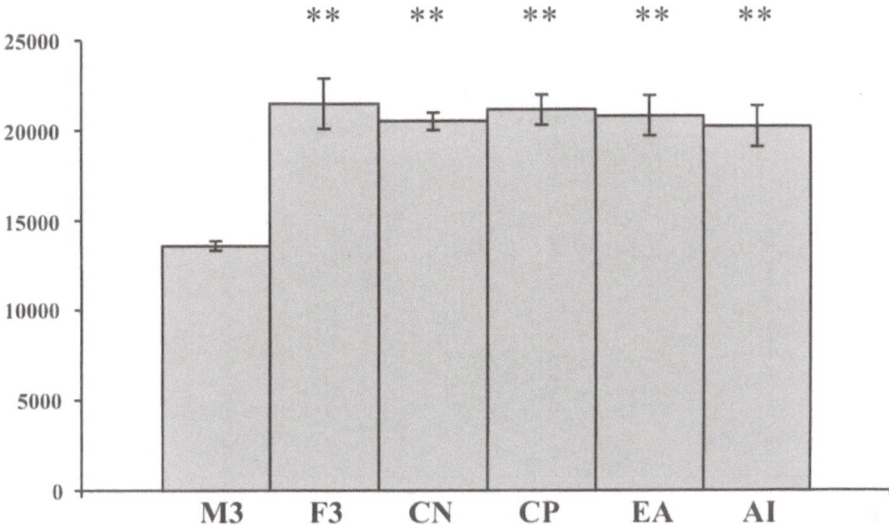

Fig. 20. Effect of manipulations of hormonal environment on the density of GABA-immunoreactive neurons in the cortical amygdala (*CoA*). Data are mean±SEM. ** $P<0.001$ compared to M3. *M3*, intact male, 3 months old; *F3*, female, 3 months old; *CN*, 3-month-old male, castrated on first day of life; *CP*, 3-month-old male, castrated at puberty; *EA*, 3-month-old male, treated with estrogen antagonist in the first 10 days of life; *AI*, 3-month-old male, treated with aromatase inhibitor in the first 10 days of life

same effect, i.e., the density of GABA-immunoreactive neurons in the MeA was higher than in intact males and lower than in females at 3 months of age. Analogous results were found for male rats that were injected with estrogen antagonist or aromatase inhibitor during the first 10 days of life (Fig. 19).

Castration of newborn as well as pubertal male rats caused erasing of the sexual dimorphism in the density of GABA-immunoreactive neurons in the CoA. Castrated males had a significantly higher density of GABA-immunoreactive neurons in the CoA than intact males at 3 months of age and did not differ from females. Similar results were observed when comparing intact males to ones that were injected with the estrogen antagonist or the aromatase inhibitor during the critical period of sexual differentiation of the brain (Fig. 20).

The density of GABA-immunoreactive neurons in the CeA was not affected by either of the experimental manipulations. Neither the castration of males nor treatment with the estrogen antagonist or the aromatase inhibitor had a significant effect in comparison to intact males at 3 months of age (Fig. 21).

Another pattern of effects was observed in the BLA after experimental manipulation of the hormonal environment The density of GABA-immunoreactive neurons was increased at 3 months of age as a result of castration of newborn and pubertal males compared to intact males of the same age. On the other hand, treatment with

CeA

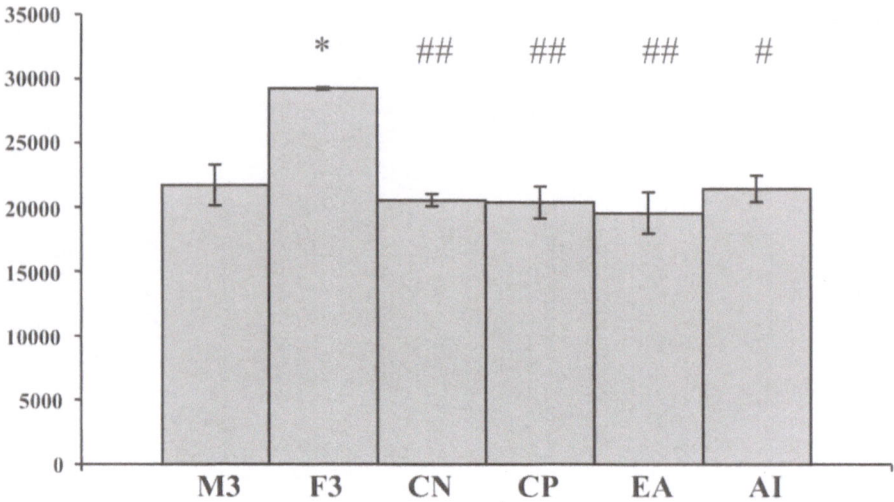

Fig. 21. Effect of manipulations of hormonal environment on the density of GABA-immunoreactive neurons in the central amygdala (*CeA*). Data are mean±SEM. *$P<0.001$ compared to M3; #$P<0.01$ or ##$P<0.001$ compared to F3. *M3*, intact male, 3 months old; *F3*, female, 3 months old; *CN*, 3-month-old male, castrated on first day of life; *CP*, 3-month-old male, castrated at puberty; *EA*, 3-month-old male, treated with estrogen antagonist in the first 10 days of life; *AI*, 3-month-old male, treated with aromatase inhibitor in the first 10 days of life

the estrogen antagonist or the aromatase inhibitor did not affect this density in the BLA (Fig. 22).

As we followed the density of GABA-immunoreactive neurons in the MeA, CoA, CeA, and BLA in four age groups, we observed that it decreased with ageing in both sexes (Table 14).

3.2.3
Leucine-Enkephalin-Immunoreactive Neurons

Leucine-enkephalin-immunoreactive neurons were observed in all the studied amygdaloid nuclei. These were predominantly small or medium-sized cells (Fig. 23). The long axis of the leucine-enkephalin-immunoreactive perikarya was measured. In the MeA it was 15.6±2.88 µm, in the CoA 14.9±2.18 µm, in the CeA 14.6±2.32 µm, and in the BLA 18.3±4.48 µm.

Morphometric studies were carried out to compare the density of leucine-enkephalin-immunoreactive neurons in the amygdala of male and female rats. At 6 days of age no sex differences were observed in the MeA, CoA, CeA, and BLA. After

BLA

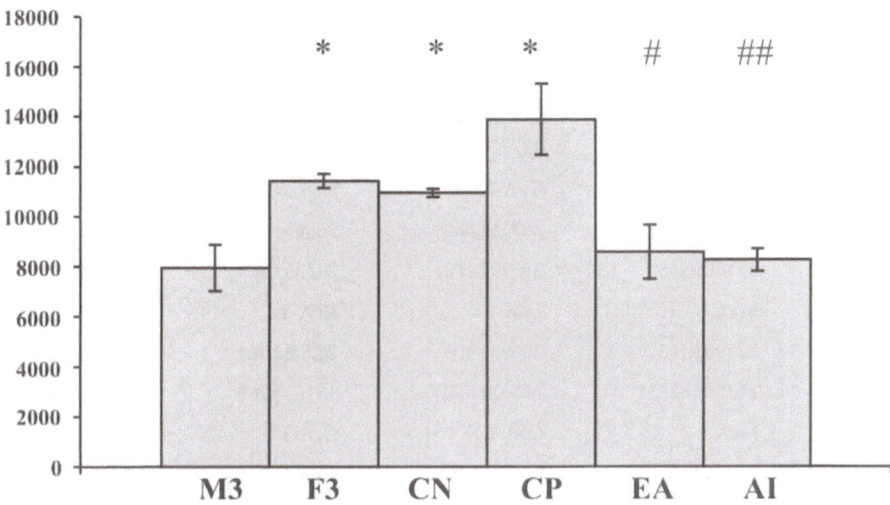

Fig. 22. Effect of manipulations of hormonal environment on the density of GABA-immunoreactive neurons in the basolateral amygdala (*BLA*). Data are mean±SEM. *$P<0.01$ compared to M3; #$P<0.01$ or ##$P<0.001$ compared to F3. *M3*, intact male, 3 months old; *F3*, female, 3 months old; *CN*, 3-month-old male, castrated on first day of life; *CP*, 3-month-old male, castrated at puberty; *EA*, 3-month-old male, treated with estrogen antagonist in the first 10 days of life; *AI*, 3-month-old male, treated with aromatase inhibitor in the first 10 days of life

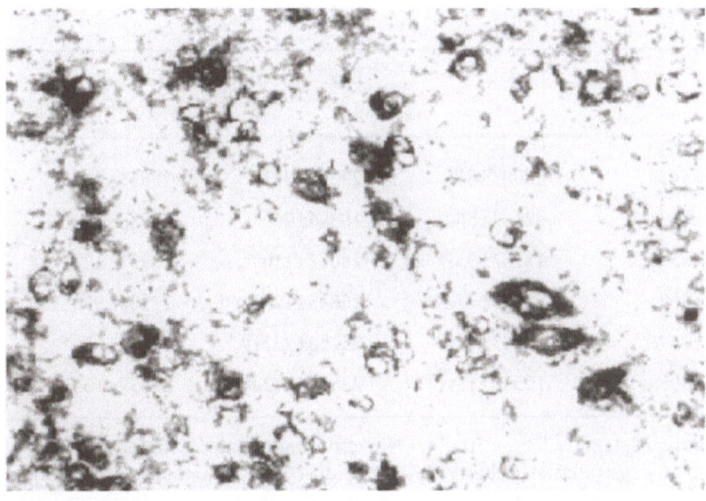

Fig. 23. Leucine-enkephalin-immunoreactive neurons in the basolateral amygdala. ×880

Table 15. Density of leucine-enkephalin-immunoreactive neurons in the medial (*MeA*), cortical (*CoA*), basolateral (*BLA*), and central (*CeA*) amygdala of male and female rats at 6 days (6), 20 days (20), 3 months (3), and 1 year (1) of age. Data are mean±SD

Leucine -enkephalin	MeA-6	CoA-6	CeA-6	BLA-6
Male	35500±901	25316±1969	30833±381	29008±605
Female	34416±1626	22775±1654	30216±503	28633±664
	MeA-20	CoA-20	CeA-20	BLA-20
Male	26966±202	25366±889	26083±381	29800±392
Female	25166±946	23550±1783	24833±763	28833±877
	MeA-3	CoA-3	CeA-3	BLA-3
Male	22896±443	26083±803	22208±381	28833±392
Female	16354±852**	22167±560*	18354±969*	23958±344**
	MeA-1	CoA-1	CeA-1	BLA-1
Male	17771±1016	22416±832	16562±286	21333±629
Female	12812±472*	17083±520**	13416±563**	17500±737*

*Sex difference at $P<0.01$; **sex difference at $P<0.001$.

Table 16. Effect of manipulations of hormonal environment on the density of GABA-immunoreactive neurons in the medial (*MeA*), cortical (*CoA*), central (*CeA*), and basolateral amygdala (*BLA*). Data are mean±SEM

Leucine -enkephalin	MeA	CoA	CeA	BLA
M3	22896±443	26083±803	22208±381	28833±392
F3	16354±852**	22167±560*	18354±969*	23958±344**
CN	13512±2502**	11387±951**##	18037±1109*	18931±1299**#
CP	18615±862*	20900±1724*	16843±1772*	22200±998**
EA	25793±3496#	22543±676*	25750±2371#	24887±851*
AI	20943±808##	19531±1210**	22528±542##	24859±1325*

M3, intact male, 3 months old; *F3*, female, 3 months old; *CN*, 3-month-old male, castrated on first day of life; *CP*, 3-month-old male, castrated at puberty; *EA*, 3-month-old male, treated with estrogen antagonist in the first 10 days of life; *AI*, 3-month-old male, treated with aromatase inhibitor in the first 10 days of life.
*$P<0.01$ or **$P<0.001$ compared to M3; #$P<0.01$ or ##$P<0.001$ compared to F3.

the end of the critical period for sexual differentiation of the brain, on the 20th day of life, still no sex differences were observed. After puberty, at 3 months and 1 year of age, the density of leucine-enkephalin-immunoreactive neurons differed between males and females in the MeA, CoA, CeA, and BLA. Males had a greater density of perikarya expressing leucine-enkephalin than females of the same age (Table 15).

Experimental manipulations of the hormonal environment during the critical period of sexual differentiation as well as in puberty were undertaken to establish whether sex differences are dependent on sex steroid levels in these life periods. Castration of newborn males caused a significant decrease in the density of leucine-enkephalin-immunoreactive neurons in the third month of life in the four studied amygdaloid nuclei compared to intact males (Table 16). In the MeA and CeA the density of leucine-enkephalin-immunoreactive neurons remained higher than in females, while in the CoA and BLA it appeared to be significantly lower even compared to female levels. Castration of males at puberty also caused a significant decrease in the density of leucine-enkephalin-immunoreactive neurons compared to intact males at 3 months of age in the MeA, CoA, CeA, and BLA. Yet, these levels did not reach female density (Table 16). Injections of the estrogen antagonist or the aromatase inhibitor to block the possible effect of aromatizable androgens in the critical period of sexual differentiation had no effect on the density of leucine-enkephalin-immunoreactive neurons in the MeA and CeA. They caused a significant difference only in the CoA and BLA in treated males compared to intact ones at 3 months of age (Table 16). In these nuclei the density of leucine-enkephalin-immunoreactive neurons decreased, but did not reach female levels.

Age-related changes in the density of leucine-enkephalin-immunoreactive neurons were established by studying four age groups of male and female rats. It appeared that the density of leucine-enkephalin-immunoreactive neurons in the MeA, CoA, CeA, and BLA smoothly decreased with ageing in both sexes (Table 15).

3.2.4
Parvalbumin-Immunoreactive Neurons

At 6 days of age, no parvalbumin-immunoreactive neuronal perikarya were found in the amygdaloid complex. In the next ages studied, three main types of parvalbumin-immunoreactive neurons were observed in the amygdaloid complex: small, oval cells; small to medium-sized neurons with multipolar perikarya; and well-marked dendritic trees and fusiform cell bodies of medium size. The long axis of the parvalbumin-immunoreactive perikarya in the MeA was 15.0±2.48 μm, in the CoA 16.1±2.04 μm, and in the BLA 21.6±2.74 μm.

The distribution of parvalbumin-positive cell bodies was unequal among the nuclei of the amygdaloid complex. In the CeA no parvalbumin-immunoreactive neurons were detected in any of the ages studied. There were also no labeled fibers in the CeA. The BLA was the richest in parvalbumin-immunoreactivity. Most of the cell bodies were multipolar medium sized with a well-visible dendritic tree (Fig. 24). Among them, single small oval neurons could be seen. Not only cell bodies but also the neuropil of the BLA showed intense parvalbumin-immunoreactivity, which clearly demarcated it from the surrounding structures. The CoA had a comparatively rare population of parvalbumin-immunoreactive neurons, which were usually small and

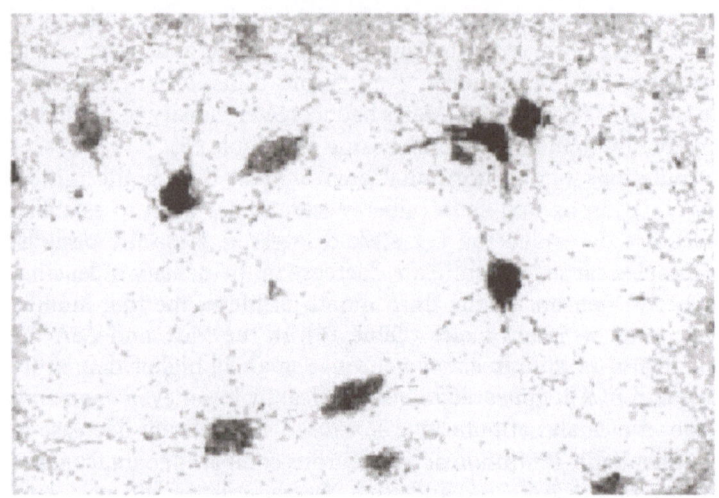

Fig. 24. Parvalbumin-immunoreactive neurons in the basolateral amygdala. ×880

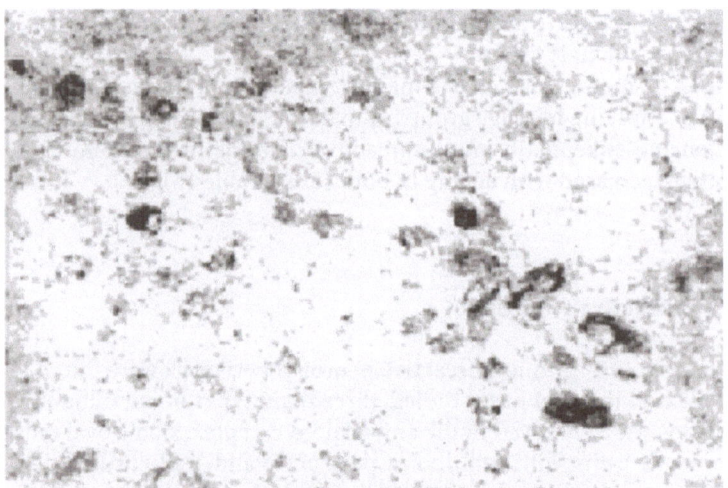

Fig. 25. Parvalbumin-immunoreactive neurons in the medial amygdala. ×880

oval. In the rostral end of the CoA a dense net of parvalbumin-positive fibers could be seen, which disappeared in more caudal levels. The parvalbumin-immunoreactive neurons in the MeA were clustered in the rostral pole (Fig. 25), while caudally they were rarely observed. In the rostral end of the MeA, parvalbumin-positive fibers could also be seen in a cluster that did not appear caudally.

Morphometric studies for comparison of the density of parvalbumin-immunore-active neurons in male and female rats were done in the MeA, CoA, and BLA. Sex

Table 17. Density of parvalbumin-immunoreactive neurons in the medial (*MeA*), cortical (*CoA*), and basolateral (*BLA*) amygdala of intact male and female rats at 6 days, 20 days, 3 months, and 1 year of age, as well as castrated newborn (*CN*) males, castrated males at puberty (*CP*), males treated with estrogen antagonist in the first 10 days of life (*EA*), and males treated with aromatase inhibitor in the first 10 days of life (*AI*), sacrificed at 3 months of age. Data are mean±SD

Parvalbumin-immuno-reactive neurons	MeA		CoA		BLA	
	Male	Female	Male	Female	Male	Female
6 days	0	0	0	0	0	0
20 days	4226±151	4091±173	6466±204	5454±138*	9508±413	9031±48
3 months	1596±131	1595±227	3975±175	2991±126**	3283±161	2802±100
1 year	1325±108	1325±109	1883±76	1012±33**	2317±104	1975±175
CN	2000±200		2558±52**		3392±128	
CP	2016±104		2717±104**		3389±130	
EA	1416±104		2226±110**		3616±104	
AI	1600±100		2108±15**		3108±210	

**P<0.001 compared to male of the corresponding age.

differences were not observed in the density of these neurons in the MeA and BLA of prepubertal (20 days old) and postpubertal (3 months and 1 year old) rats. In the CoA, sexual dimorphism was found at 20 days of age. Males had a greater density of parvalbumin-immunoreactive neurons than females of the same age. This difference was preserved and was also expressed in postpubertal animals (Table 17).

Experimental manipulations of the androgen levels and their possible effect on the expression of parvalbumin in the MeA, CoA, and BLA were performed. They significantly affected only the parvalbumin-immunoreactive neuronal population in the CoA. Castrations of newborn males and pubertal males, as well as treatment with estrogen antagonist or aromatase inhibitor in the first 10 days of postnatal life, caused a decrease in the density of these neurons in comparison to intact males at the age of 3 months (Table 17).

Age-related changes in the density of parvalbumin-immunoreactive neurons were followed. It was established that parvalbumin-positive neurons in the MeA, CoA, and BLA appear between the 6th and 20th day of postnatal life. After this age a slow decrease in the density of parvalbumin-immunoreactive cells was observed at 3 months and 1 year of age in both sexes (Table 17).

4 Discussion

The results of this study confirm the existence of sexual dimorphism in the two limbic structures studied – the bed nucleus of the stria terminalis and the amygdala. Several new aspects of sex differences are established together with the confirmation of previously reported gender-specific patterns of volume and neuronal densities in some of the nuclei. The experimental manipulations of the hormonal environment during the critical period of sexual differentiation and in puberty allow one to make some conclusions on the possible mechanisms that are occupied with the formation of sex differences in the BST and the amygdaloid nuclei. The expression of sex differences in various life periods is detected to establish the time of occurrence.

Up to now, most investigations of sexual dimorphism in the amygdala have been directed to the medial amygdala. It has a greater volume in male than in female rats (Mizukami et al. 1983; Hines et al. 1992; Cooke et al. 1999). Sex differences are reported in the expression of several neuropeptides, i.e., CCK (Micevych et al. 1988), vasopressin (De Vries and Al-Shamma 1990; Szot and Dorsa 1993; Lakhdar-Ghazal et al. 1995; Wang and De Vries 1996), substance P (Malsbury and McKay 1989), and somatostatin (Baldino et al. 1988). Nishizuka and Arai (1981) have demonstrated that the synaptic density in the medial amygdala is higher in male rats and is controlled by androgen levels during the neonatal period. Kalimullina (1985, 1986, 1988) described sex difference in the size of the neuronal nuclei in the medial, cortical, and basolateral amygdala.

Several examples for sexual dimorphism of the BST are present in the literature. Guillamón et al. (1988) found greater neuronal numbers in the male posterior medial subnucleus of the BST, while in the anterior medial subnucleus, female rats have a higher neuronal density than males. The volume of the posterior medial subnucleus of the BST appears to be greater in male rats (Del Abril et al. 1987) and male guinea pigs (Hines et al. 1985). Sex differences are established in the distribution of several neurotransmitters: CCK (Micevych et al. 1988), vasopressin (Van Leeuwen et al. 1985; De Vries and Al Shamma 1990; Szot and Dorsa 1993; Lakhdar-Ghazal et al. 1995; Al-Shamma and De Vries 1996), and substance P (Malsbury and McKay 1987). Castration of male rats causes a decrease in the number of neurons expressing mRNA for galanin (Miller et al. 1993). Sex differences in the distribution of opioid receptors have been reported (Ostrowski et al. 1987). The BST has been shown to possess sex differences not only in its structure, but also in its projections. Hutton et al. (1998) described sexually dimorphic projections to the anteroventral periventricular nucleus of the preoptic area.

It is of great interest that sex differences in the human BST have been found and could even be related to the problem of sexual identification. Allen and Gorski (1990) have proven that the dark-staining portion of the posteromedial BST is 2.47 times larger in men than in women. In 1995, Zhou et al. found that the central portion of the BST is larger in heterosexual men than in women and transsexual men.

The BST and the amygdala are connected between themselves as well as with other sexually dimorphic structures in a sexually dimorphic circuitry, which participates in the control of reproductive behavior. According to Segovia and Guillamón (1993, 1996), the BST and amygdala are major elements of the vomeronasal pathway, directly related to the control of lordosis behavior, maternal behavior, and male sexual behaviors in rodents (Segovia and Guillamón 1993, 1996; Kondo and Yamanouchi 1995; Wood and Newman 1995). Searching for sex differences in this neural network is important for understanding the complex neurobiological response pattern in both genders. It is accepted that most sex differences in the vomeronasal system are androgen dependent.

The results of our study as well as data in the literature suggest that sexual differentiation of the CNS is a complex process. It is obvious that it would be too simplified to consider that sexual differentiation of the brain is regulated by gonadal steroids only after the proposed classical theory (Fig. 1). The generation of several sexually dimorphic features can be explained by this theory. However, there is a great amount of information on sex differences in the mammalian CNS that fails to fit the generally accepted classical theory. It appears that sex steroids are capable of influencing brain development and plasticity in various ways and diverse time spans. Gonadal steroids are a powerful plastic factor able to change the brain sexual phenotype through binding to estrogen and androgen receptors, which are present in the developing as well as in the adult brain. The gonadal steroids acting in the brain environment can affect the genome of cells and thus settle a stable sexually dimorphic characteristic. Further, sex steroids can probably affect other environmental factors, for example, growth factors and neuromediators and thus fulfil their differentiating role through secondary pathways. Late activating effects of androgens and estrogens are apparently also necessary for the phenotypic expression of certain gender specificity. Unfortunately, most of the mechanisms of these actions remain enigmatic up to now. Since the late effects of androgens and estrogens are reversible in most cases, it could be proposed that they act through other mechanisms than genomic influencing ones. It is interesting for future studies on sexual differentiation of the brain to consider that in one and the same neuronal population different characteristics, such as neuronal numbers and expression of various neurotransmitters, can be differently regulated by levels of gonadal steroids. For example, our results confirm that in the BST of the rat there is no sex difference in the total neuronal density in the lateral, medial, and ventral subnuclei. At the same time we show sex difference in the density of GABA-immunoreactive neurons that are more numerous in the female than in the male BST. Further, leucine-enkephalin-immunoreactive neurons in the lateral BST have greater density in male rats. The experimental results suggest that GABA expression in the subdivisions of the rat BST is regulated by androgens mostly by late postpubertal actions, while leucine-enkephalin expression in the lateral BST seems to be determined during the critical perinatal period for sexual differentiation of the brain but not later in postpubertal life. Similarly, we have reported different mechanisms of sexual differentiation in the medial, cortical, central, and basolateral amygdaloid nuclei. These facts additionally support the notion that androgens and estrogens act in several convergent ways to modify the final sexually specific phenotype of the brain structure and its various attributes. Whether this effect is associated with direct action of the sex steroids on the examined cellular population or is due to secondary cell–cell interaction and modification should also be taken into consideration when evaluating sexual differentiation of the CNS.

In addition to the role of gonadal steroids for the sexual differentiation of the brain, it is important to expand our knowledge on the genetic mechanisms that underlie this process. Since the Sry gene is transcribed in the male mouse and human brain (Lahr et al. 1995; Mayer et al. 1998), it seems that general genetic factors also modify the gender characteristics of the brain irrespective of the hormonal environment. However, autosome and gonosome genes which could affect sex-specific behaviors and brain morphology might be additionally assisted in their sexually differentiating action by androgens and estrogens or other environmental factors. Present information suggests much richer and complicated interactions of factors and mechanisms that make it difficult to evaluate their contribution to the sexual differentiation of the CNS.

4.1
Localization of the Nuclei

In the present study the BST and its three main subdivisions – medial, lateral, and ventral – are described on Nissl-stained preparations. They correspond to the medial, lateral, and ventral part of the BST as described by Moga et al. (1989). As these authors point out, each of the parts of the BST can be divided into several subnuclei according to their heterogeneous chemoarchitecture and projections. To avoid inaccuracy in quantitative examinations, counting of neurons in the subdivisions of the BST is done by equal distribution of the morphometrically studied images in all the subnuclei from medial to lateral and from rostral to caudal parts.

The amygdala is classically divided into a basolateral group of nuclei and deep amygdaloid nuclei (Krettek and Price 1978a). In our study we have analyzed the basolateral and central amygdaloid nuclei as representatives of the basolateral group, and the cortical and medial amygdaloid nuclei as representatives of the deep nuclei. The basolateral, central, and medial amygdala that are described in the present study correspond exactly to the homonymous nuclei by Krettek and Price (1978a). The cortical amygdala described here corresponds exactly to the localization of the anterior CeA in the rat as described by Krettek and Price (1978a). Again, to avoid errors in morphometric studies because of the heterogeneity of the amygdaloid nuclei, all images analyzed are equally distributed among the subdivisions medially to laterally and from the rostral to caudal extent of each nucleus.

4.2
Volumes

Our data on the volumes of the BST in 6-day-, 20-day-, 3-month-, and 1-year-old rats revealed no sex difference in any of the ages studied. These results confirmed data of Del Abril et al. (1987), who have shown that there is no sex difference in the volume of the BST of 3-month-old Wistar rats. Additionally, we proved that the volume of the

BST is not different between males and females before the end of the critical period of sexual differentiation of the brain, in prepubertal and ageing rats. A gradual increase of the BST volume is observed while reaching the volume of adult animals between 20 days and 3 months of age, i.e., around puberty. Our results show that androgens during the critical period of sexual differentiation and in puberty do not affect the total volume of the BST. Data on castration of newborn male rats confirm the results of Del Abril et al. (1987) that it has no effect on the volume of the BST. We also prove that treatment with the estrogen antagonist or the aromatase inhibitor during the same period does not influence the volume of the BST. When summarizing these data, we confirm that sex steroids do not have sex-specific determining effects on the BST volume during the critical period of sexual differentiation of the developing rat brain. Castration of male rats in puberty caused no changes concerning the volume of the BST, which proves that postpubertal levels of sex steroids do not affect the volume of the BST either. Del Abril et al. (1987) have found sex differences in certain subdivisions of the BST but because of their opposite direction they do not have any influence on the total volume of the nucleus. These authors have found that the rostral part of the medial BST has greater volume in females, while the caudal part is larger in males. Del Abril et al. (1987) report that castration of newborn males caused a decrease in the volume of the caudal part and an increase in the volume of the rostral part of the medial BST without influencing the total volume of the medial BST. Androgenization of newborn females, on the other hand, causes an increase in the volume of the posterior medial region, but does not affect the volume of the anterior medial region of the BST. Similar results have been reported also in the medial BST of guinea pigs (Hines et al. 1985). Obviously, sex steroids control the volume of the posterior medial part of the BST during the critical period of sexual differentiation of the developing brain. It does not reflect the total volume of the nucleus since there are other parts with opposite correlations between sexes whose control by sex steroids is more difficult to explain with the existing data.

The volume of the medial, cortical, central, and basolateral amygdala is variously affected by gender. Basolateral, cortical, and central amygdaloid nuclei have not shown sex differences in their volume in any of the age groups studied. No effect of castrations of newborn or pubertal males has been detected on the volume of these nuclei. Experimental treatment with the estrogen antagonist or the aromatase inhibitor during the first 10 days of life confirms that androgens do not affect the volume of the basolateral, central, and cortical amygdala of rats. This is not valid for the medial amygdala. Our data confirm that the volume of the medial amygdala is strongly dependent on sex. Sex difference is detected on the 20th day of life for the first time. It is present in all the subsequent postpubertal ages studied. Males have a greater volume of medial amygdala than females. These data correspond to the results of Mizukami et al. (1983), who have found the same type of sex difference in postnatal rats. Using estradiol treatment of female rats, Mizukami et al. (1983) proved that the sexual differentiation of the volume of the medial amygdala is accomplished during the early stages of postnatal development under the organizational effect of sex steroids. Our original results from the experimental castration of newborn male rats, as well as treatment with the estrogen antagonist or the aromatase inhibitor, confirm the idea that the volume of the medial amygdala is sexually differentiated during the critical period early after birth. The absence of androgens during this period or blocking of their aromatization to estrogen and acting on the neuronal genome have a unidirectional effect, i.e., decrease of the male nuclear volume to female levels. Therefore, sexual differentiation of the volume of the medial amygdala is fulfilled in accordance to the classical theory (Pilgrim 1994) for sexual differentiation of the brain. Sexual dimorphism in the volume of the MeA is under the epigenetic control of gonadal steroids during the early critical period of sexual differentiation and not later when androgens could have activational effects. Recently it has been demonstrated that the volume of the posterodorsal part of the MeA is sensitive to androgens in adult rats too (Cooke et al. 1999). Although we observed some decrease in the total volume of the MeA after postpubertal castration of males, it was not statistically significant. Most probably, there are subregions and subpopulations of neurons within the MeA, as it has also been demonstrated in the BST, that are regulated differently by gonadal steroids.

4.3
Total Neuronal Density

The other morphological characteristic that we have studied on cresyl-violet-stained sections is the total neuronal density in the BST and the four amygdaloid nuclei. Our results show that there is no sex difference in the neuronal density of the lateral, medial, and ventral part of the BST in any of the age groups studied. Experimental castration of newborn and pubertal males as well as treatment with the estrogen antagonist or the aromatase inhibitor have no effect on the neuronal density of the lateral and ventral BST. In the medial BST, androgen deprivation on the first day of life and treatment with the estrogen antagonist or the aromatase inhibitor of male rats during the critical period for sexual differentiation cause a decrease in the neuronal density in comparison to intact males at the age of 3 months. An explanation of this result can be found in the data reported previously by Guillamón et al. (1988) in 3-month-old Wistar rats. These authors found that in the posterior medial region of the male BST there is a significantly higher number of neurons than in females. They established that a population of neurons in the posterior medial BST is sensitive to androgen levels in the days immediately after birth. Androgen presence is critical for the expression of sexual dimorphism in this part of the nucleus. This androgen-sensitive neuronal group is probably responsible for the decrease of the neuronal density in the medial BST as reported in the present study. On the other hand, the anterior lateral region of the BST has greater neuronal density in females than in male rats (Guillamón et al. 1988), but this sexual dimorphism could not be explained by the action of aromatizable androgens during the early critical period for sexual differentiation of the brain. This idea is supported by our results, since the experimental influencing of the androgen action through blocking aromatase activity or estrogen receptors does not affect the neuronal density of the lateral BST.

In the amygdala, various effects of gender and gonadal steroids upon the neuronal density of different amygdaloid nuclei have been detected. The basolateral and central amygdaloid nuclei, which belong to the deep amygdaloid nuclear group, have not shown any sex difference in their neuronal density, neither in prepubertal 6- and 20-day-old rats nor in postpubertal 3-month- and 1-year-old rats. The androgen levels do not affect the neuronal density in these nuclei in the neonatal period or after puberty. Treatment with the estrogen antagonist or the aromatase inhibitor during the first ten postnatal days has no effect on neuronal density in the basolateral and central amygdala. Therefore, we can conclude that neuronal density in the basolateral and central amygdala is not affected by the sexually differentiating effect of gonadal steroids.

In the medial and cortical amygdala of 6-day-old rats, no sex difference is found concerning their neuronal density. But in the next age groups it is established that males have greater neuronal density than females. The experimental manipulations of the hormonal environment support the idea that aromatizable androgens during the critical period for sexual differentiation of the brain are responsible for the formation of the sex differences. Deprivation of androgens immediately after birth due to castration of newborn males and treatment with the aromatase inhibitor or the estrogen antagonist, which block the aromatization of androgens or their action on the neural genome through estrogen receptors, have unidirectional effects. Each of the experimental groups mentioned have lower neuronal density in the MeA and CoA than intact males in the third month of life. On the other hand, the late decrease of androgen levels by castration in puberty does not affect the neuronal density in the MeA and CoA. Supporting our conclusions are the data on distribution of estrogen receptors in the amygdala. Simerly et al. (1990) and Shughrue et al. (1997a) show that the highest expression of estrogen receptors is found in the MeA and CoA. Additionally, the aromatase activity in these nuclei is very high (Lephart 1996), which proves the idea that aromatizable androgens play a major role in the formation of sexual dimorphism in the medial and cortical amygdaloid nuclei.

Our data for higher neuronal density in the MeA and CoA in male animals and greater volume of the MeA in males are similar to the results of other investigations in various brain regions with sexual dimorphism: the sexually dimorphic nucleus of the preoptic area (Gorski et al. 1980), the anterior hypothalamus (Allen et al. 1989; LeVay 1991, 1993), the visual cortex (Reid and Juraska 1992), Onuf's nucleus and its analogues (Breedlove and Arnold 1980; Forger et al. 1996), and the hippocampus (Wimer and Wimer 1985) are found to have greater neuronal densities and greater volumes in male than female rats.

4.4
GABA-Immunoreactive Neurons

Our results reveal the presence of sex difference in the GABAergic neuronal population in the BST and amygdala (Stefanova et al. 1997a, 1998; Stefanova 1998), and for the first time we establish the mechanisms for sexual differentiation of these neurons. In the lateral and ventral BST the density of GABA-immunoreactive neurons is sexually dimorphic only in postpubertal rats. In all the cases, females have a greater density of GABA-immunoreactive neurons than males. Data from the experimental manipulations of the hormonal environment explain to some extent the mechanism for generation of this sex difference. Castrations on the first postnatal day and in puberty lead to an increase in the density of GABA-immunoreactive neurons in the lateral and ventral BST. On the other hand, treatment with the estrogen antagonist and the aromatase inhibitor remains without any effect. Our results correspond to data of Lieb et al. (1994), who found no sexual differentiation of prenatal GABAergic neurons in vitro under the influence of estrogen or testosterone. Since sex differences appearing after puberty and castrations, i.e., lack of androgens, delete this difference, we can conclude that androgens influence the sexual differentiation of GABAergic neurons in the lateral and ventral BST after puberty. This is not in accordance to the classical theory for sexual differentiation in the critical perinatal period. It is probable that androgens downregulate the expression of GABA in the neurons of males after puberty through their action on their receptors, which are numerous in the BST (Simerly et al. 1990). Grattan and Selmanoff (1993) presented data on the influence of testosterone on the GABAergic system in hypothalamic nuclei. They found that following castration of male rats, the steady-state GABA concentrations increase and GABA turnover decreases in the diagonal band of Broca, the medial preoptic area, and the median eminence. It is probable that the mechanisms of generating sex difference in the lateral and ventral BST are complex and stimulatory effects of estrogens contribute to the sexual dimorphism of GABAergic neurons. This idea is suggested by the results of McCarthy et al. (1995), who have established that estrogens can influence the expression of GAD.

A different pattern of correlation between androgens and expression of GABA is found in the medial BST. In this part of the nucleus sex differences are established prepubertally at 20 days of age and stay stable in the postpubertal ages. On the other hand, all experimental manipulations of the hormonal environment, i.e., castrations of newborn and pubertal males as well as treatment with estrogen antagonist or aromatase inhibitor, have a unidirectional effect. This concerns the increase of the density of GABA-immunoreactive neurons in the medial BST compared to intact males in the third month of life. In this part of the BST other factors take part in the sexual differentiation of the GABA-immunoreactive neurons, obviously together with the mechanism of testosterone influence discussed above. The medial BST has been shown to have high concentrations of estrogen receptors (Simerly et al. 1990). As it has been found in Nissl-stained preparations, the medial BST possesses a group of neurons whose density is sexually differentiated according to the classical theory for influence of aromatizable androgens. Our results prove that the expression of GABA in the medial BST is determined during the perinatal period in the presence of aromatizable androgens. As Pilgrim and Hutchison (1994) point out in their review, sex differences in some peptide-ergic neurons, which are settled during the perinatal period, undergo additional modification in later stages of individual life. The control of GABA expression in the neurons of the medial BST is probably also under additional modification, as it is suggested by the results we observed in castrated pubertal rats.

The idea of such a mechanism for sexual differentiation of GABA-immunoreactive neurons is supported by our results on density of GABA-immunoreactive neurons in the amygdaloid nuclei (Stefanova 1998). In the basolateral amygdala, similar to the results in the lateral and ventral BST, sex difference in the density of GABA-immunoreactive neurons is observed after puberty. Females have a greater density than males. This sex difference is affected by castrations of newborn or pubertal male rats. Castrated animals have a greater density of GABA-immunoreactive neurons than intact males in the third month of life and this density is comparable to the density in female rats of the same age. Treatment of newborn males with the estrogen antagonist or the aromatase inhibitor during the first 10 days of life has no effect on the density of GABA-immunoreactive neurons. These data, as well as the information on the presence of androgen receptors in the BLA (Simerly et al. 1990), suggest that in this nucleus, a ndrogens and probably the interaction with their receptors after puberty are a major factor for the formation of the noted sex differences. However, other indirect interactions can not be excluded.

The results of the medial and cortical amygdala are similar to those reported in the medial BST. Sex differences are observed before puberty at 20 days of age and are preserved in postpubertal rats. Females have a greater density of GABA-immunoreactive neurons than males. The medial and cortical amygdala possess a high concentration of estrogen receptors (Simerly et al. 1990; Shughrue et al. 1997a) and high aromatase activity (Lephart 1996). As a result of castration and treatment with the estrogen antagonist or the aromatase inhibitor of newborn males, an increase in the density of GABA-immunoreactive neurons in the medial and cortical amygdala is established as compared to intact males in the third month of life. It is suggested that sexual differentiation of GABA-containing nerve cells in these amygdaloid nuclei is accomplished during the perinatal critical period for sexual differentiation of the brain under the influence of aromatizable androgens and interaction with estrogen receptors. This is probably not the only mechanism by which sex steroids fulfil their sexually differentiating effect, since castration of males in puberty also causes an increase in the density of GABA-immunoreactive neurons in the medial and cortical amygdala. Additionally, treatment with the estrogen antagonist or the aromatase inhibitor gives no female levels in the density of GABA-immunoreactive neurons as after castration of newborn male rats. Obviously, other late activational mechanisms of androgen modulate the expression of GABA in these amygdaloid nuclei. This mechanism could be the interaction with androgen receptors (Simerly et al. 1990) and induction of the genome regulation of the GABA metabolism.

Results in the central amygdala differ significantly from those in the other amygdaloid nuclei studied. Sexual dimorphism in the density of GABA-immunoreactive neurons is found in postpubertal rats. Females have a greater density of GABA-immunoreactive neurons than males. Neither of the experiments performed has affected the sexual dimorphism of GABA-containing neurons in the central amygdala. Obviously, androgens are not the main factor for sexual differentiation of this neuronal population, although high concentrations of androgen receptors are established in the central amygdala (Simerly et al. 1990). Several explanations of these results can be suggested. It is possible that sexual dimorphism in the density of GABA-immunoreactive neurons is genetically predetermined. However, its expression after puberty suggests that there is some connection with the change of the gonadal steroid environment during this period. It cannot be excluded that estrogens play such a role, since estrogen receptors are found in the central amygdala (Simerly et al. 1990; Shughrue et al. 1997a). Another possibility is that sex steroids have a regulatory effect on another brain structure connected to the central amygdala [for instance the preoptic area (Russhen 1982b)]. Pilgrim and Hutchison (1994) discuss this mechanism for sexual differentiation as cell-to-cell interactions, which mediate transsynaptically hormonal effects.

Our results for sexual dimorphism in the density of GABA-immunoreactive neurons in the BST and the amygdala (Stefanova 1998; Stefanova et al. 1997a, 1998) correspond to data on sex differences of GABAergic systems in other brain regions, which are also related to the control of reproductive behavior. These are sex differences in the tissue concentrations of GABA (Frankfurt et al. 1984), the GABA-A receptor binding (Jüptner and Hiemke 1990), and the changes in GABA receptors by gonadal steroids (Lasaga et al. 1988). The sex difference of GABA expression in the BST and the amygdala is variously settled in the different regions. In all cases of sexual dimorphism, females have greater density of GABA-immunoreactive neurons than males. These results correspond to results that are reported in the prenatal striatum (Ovtscharoff et al. 1992) for prevalence of GABA-immunoreactive neuron densities in female fetuses. Similar data have been presented in 20-day-old rat striatum (Ovtscharoff et al. 1997; Bozhilova-Pastirova and Ovtscharoff 1996). The control of sex difference generation is complex and includes the direct influence of gonadal steroids through their receptors, as well as indirect mechanisms of intercellular interactions. Together with the late activational effects after puberty, sex hormones perform the possible organizational effects on the expression of GABA in the BST and the amygdala during the perinatal period of life. The higher expressions of GABA in higher amounts of neurons in female animals, because of the positive induction of GAD by estrogens (McCarthy et al. 1995) or higher degree of cell death (Pilgrim 1994) of non-GABAergic neurons in males, could be suggested as putative mechanisms for sexual differentiation.

The sex differences in the density of GABA-immunoreactive neurons in the BST and the amygdala of different age groups in this study demonstrate interesting age-related changes. In the amygdaloid nuclei the density of GABA-immunoreactive neurons decreased with ageing in male as well as female rats. This result is in good agreement with previously reported data by Lolova and Davidoff (1991). This is comprehensible, since the total neuronal density also decreases with ageing of the amygdala. However, despite the decrease of total neuronal density in the BST, the density of GABA-immunoreactive neurons in the main parts of the BST contradictably increases with ageing (Stefanova et al. 1998). In both genders this increase has the same course. This fact suggests that other factors, and not gonadal steroids, determine this process, which is especially well expressed after puberty. It is possible that the higher density of GABA-immunoreactive neurons in older rats is due to the higher cell death rate of non-GABA-expressing neurons. Yet, the increase is not only proportional but can be found in the absolute values of GABA-immunoreactive neuronal densities. Therefore, it is more probable that a higher expression of GABA in the neurons of the BST is present in older animals. It could be suggested that these additionally arising GABA-immunoreactive neurons have not been detected in younger ages, because the GABA content has not been high enough to mark the cell. Another possibility is the induction of enzyme activity with ageing of enzymes, which participate in synthesis or degradation of GABA and thus elevate the intracellular content of the neurotransmitter to be elevated. Taking into consideration the results of Letinic and Kostovic (1996), transitional expression of various markers in the same neuronal population could be supposed. Micheva and Beaulieu (1995) found that the expression of cortical GABA might reflect delayed maturation and adaptation of this inhibitory circuitry. These authors suggested the existence of three subpopulations of GABA neurons depending on the time of onset and the regulation of GABA expression. There are neurons which express GABA later in development and represent a substrate of experience-dependent plasticity, which could be the case also in the BST. Recently, Fortin et al. (1999) demonstrated segmental specification of GABAergic inhibition during the development of hindbrain neural networks that could determine later maturation of GABAergic inhibition. Data that continuous neurogenesis is present in the hippocampus (Gould et al. 1998) could suggest, with reserve, such a mechanism for elevation of the density of GABA-expressing neurons in the BST. Methods used in the present study do not have the power to determine exactly the factors and mechanisms which control the increase in density of GABA-immunoreactive neurons in the BST. The data present in the literature allow speculation on the hypothesis of estrogen induction of GABA synthesis in this nucleus. McCarthy et al. (1995) prove the influence of estrogens on the expression of GAD mRNA in some hypothalamic nuclei. Estrogens after puberty may be the factor for activation of GAD mRNA, hence elevated synthesis of GABA in female BST. In males the same effect could be gained by aromatization of androgens to estradiol, since in the BST, in contrast to other brain regions, high aromatase activity is preserved in adults (Shinoda 1994; Beyer and Hutchison 1997). However, if this is the mechanism for higher expression of GABA, it means that it acts regionally specific, because no such effect is found in the amygdaloid nuclei where a high concentration of estrogen receptors and high aromatase activity are also found in adults.

4.5
Parvalbumin-Immunoreactive Neurons

Parvalbumin-immunoreactive neurons are in most cases considered a subpopulation of GABAergic neurons (Freund 1989; Cowan et al. 1990). Yet, there are data that parvalbumin-containing neurons which are GABA-negative exist (Polgar and Antal 1995). Up to now, no description of the distribution of parvalbumin-immunoreactive neurons in the BST has been presented. Our results show that despite the rich GABA-immunoreactive population in the BST, the expression of the calcium-binding protein parvalbumin is poor. Single parvalbumin-immunoreactive neuronal perikarya could be observed only in the ventrolateral region of the BST of the rat. These cell bodies are small with poorly marked beginnings of dendrites. These are most probably interneurons that sprout inside the BST and form the local circuitry. Our morphometric studies show no sexual dimorphism in the density of parvalbumin-immunoreactive neurons in the rat BST. The experimental castrations and treatment with the estrogen antagonist or the aromatase inhibitor prove that parvalbumin-containing neurons in the BST are not dependent on gonadal steroid levels. Our results confirm that parvalbumin expression in the BST appears later than the first postnatal week as it is described in other brain regions (Alcantara et al. 1993; Bozhilova-Pastirova et al. 1997a; Stefanova et al. 1997c). The highest density of parvalbumin-immunoreactive neurons is established in 20-day-old rats, while in postpubertal 3-month- and 1-year-old animals the density slowly decreases as expected because of the decrease in the total neuronal density in the BST.

Data of parvalbumin-immunocytochemical studies in the rat amygdaloid complex confirm the results of previous investigations (Celio 1990). No parvalbumin-immunoreactive neurons can be found in the rat central amygdala despite the great density of GABA-containing neurons (Celio 1990; Jolkkonen and Pitkänen 1998). In the medial amygdala a small density of parvalbumin-containing neurons is found, which suggests that they are not a major element of the inhibitory system in the control of the amygdaloid projections to endocrine and autonomic centers in the hypothalamus and brain stem. The BLA appears to be the richest in parvalbumin-immunoreactive neurons. Intensely marked perikarya with small to medium size and well-delineated dendritic trees can be seen in the basolateral amygdala. Most probably they are local interneurons. Data for the distribution of parvalbumin in the amygdaloid complex have been presented in primates and humans (Pitkänen and Amaral 1993b; Sorvari et al. 1995) and they are quite similar to those in rats.

Our morphometric studies reveal that parvalbumin-immunoreactive neurons in the cortical amygdala are dependent on gender and levels of sex steroids during critical periods of development (Stefanova et al. 1997b). Sex difference in the density of parvalbumin-containing neurons is established at 20 days of age and is preserved in postpubertal rats. Experimental castration of newborn males and blocking of estrogen receptors or inhibition of aromatase enzyme during the perinatal critical period cause a decrease in the density of parvalbumin-immunoreactive neurons in the cortical amygdala. It is lower than intact males and comparable to female levels. These results prove that androgens have an organizational effect in the perinatal period through aromatization to estrogens, and form the male type of parvalbumin expression in the rat cortical amygdala. Obviously, the effect of androgens is not limited only to this period, since a change in the density of parvalbumin-immunoreactive neurons is also observed after castration of males in puberty. Therefore, androgens also have late activational effects on this neuronal type. It is possible that androgens perform some protective effect on neurons expressing parvalbumin, which may be the reason for the appearance of sex difference. Other brain areas have also been reported to show sexual dimorphism concerning parvalbumin-immunoreactive neurons (Bozhilova-Pastirova and Ovtscharoff 1996; Stefanova et al. 1997b; Karamfiloff et al. 1998).

4.6
Leucine-Enkephalin-Immunoreactive Neurons

The results of our morphometric studies reveal the existence of sex difference in the density of leucine-enkephalin-immunoreactive neurons in the lateral BST. Females in all the age groups studied have lower density than males. This sexual dimorphism is expressed even in the youngest age studied – 6 days. Data from experimental manipulation of the hormonal environment prove the importance of androgens for the formation of sex difference of leucine-enkephalin-immunoreactive cells in the lateral BST. Castrated immediately after birth and treated with the estrogen antagonist or the aromatase inhibitor, experimental male groups have shown a significantly lower density of leucine-enkephalin-immunoreactive neurons in the lateral BST than intact males. Yet, they do not reach female levels. It is suggested that sexual dimorphism is generated in accordance to the classical theory for sexual differentiation of the brain under the influence of aromatizable androgens during the perinatal period. Testosterone, which is secreted by male gonads during the perinatal period, is aromatized in the brain to estradiol that determines the male phenotype of leucine-enkephalin expression in the lateral BST. The fact that sex difference is expressed as early as the sixth day of postnatal life, as well as the experimental results that androgens affect it partially (not reaching female levels), suggest that the formation of sexual dimorphism starts before birth. According to the present information, the mechanisms, which may regulate neuronal densities, are either sexually specific neurogenesis or programmed cell death (Pilgrim 1994). It seems that the influencing of programmed cell death and a specific neurotrophic effect of androgens in males are the most probable mechanisms for sexual differentiation of leucine-enkephalin-immunoreactive neurons in the lateral BST.

Sex differences in the density of leucine-enkephalin-immunoreactive neurons in the medial, cortical, central, and basolateral amygdala are expressed after puberty. Male rats have a greater density for these cells than females. The results from experimental manipulations of the hormonal environment reveal that the formation of sexual dimorphism of leucine-enkephalin-immunoreactive neurons in the amygdala is complex, regionally specific, and most probably regulated by several mechanisms.

In the medial and central amygdala, treatment with the estrogen antagonist or the aromatase inhibitor during the first 10 days of postnatal life does not affect the density of leucine-enkephalin-immunoreactive neurons. On the other hand, castrations of newborn or pubertal male rats cause a decrease of the density. Summarizing these results and taking into consideration that sex differences appear after puberty, it is suggested that androgens perform their sexually differentiating effect after puberty. Therefore, it is more likely to propose activational effects of androgens on the expression of leucine-enkephalin in these nuclei whose mechanism is difficult to explain with the methods used in this study.

In the cortical and basolateral amygdala, together with a decrease of the density of leucine-enkephalin-immunoreactive neurons after castrations, a similar effect is found after treatment with the estrogen antagonist or aromatase inhibitor. Obviously, together with the late activational effect, androgens also perform early organizational effects during perinatal period. As proof it is pointed out that castration of newborn males, which deprives the androgen action in the perinatal period as well as after puberty, causes a more significant decrease in the density of leucine-enkephalin-immunoreactive neurons. It is not only lower than the density in intact males, but also lower compared to females. Similar results have been presented in the hypothalamus (Simerly et al. 1988). Sexual dimorphism of enkephalin-ergic neurons is found in the anteroventral periventricular nucleus of the preoptic area. Male rats show a greater number of enkephalin-containing neurons than females. Sexual dimorphism is partially dependent on gonadal steroid levels during the perinatal period. Analogously to our results, Simerly et al. (1988) have found that castration of newborn males causes a decrease in the number of enkephalin-ergic neurons in comparison to intact males, and yet it does not reach female levels. At the same time, the same authors have not established sex difference in the number of enkephalin-ergic neurons in the anterodorsal preoptic nucleus and thus it remains unaffected by castration immediately after birth. If we summarize our results and the results of Simerly et al. (1988), we can conclude that sex steroids control the density of leucine-enkephalin-immunoreactive neurons in a different way in various brain nuclei and most probably is not limited to the perinatal period.

4.7
Sexual Dimorphism of the CNS

In recent years there has been growing evidence for the existence of sex differences in the structure of the vertebrate CNS. Some of them are directly related to the control of a sex-specific function or behavior. An especially outstanding example is the sexual

Table 18. Some examples of sexual dimorphism in the CNS

Sexually dimorphic regions of the CNS	References
Cerebrum	
Neocortex	Reid and Juraska 1992
Accessory olfactory bulb	Roos et al. 1988
Bed nucleus of the stria terminalis	Allen and Gorski 1990; Stefanova et al. 1997a,b, 1998a
Amygdala	Hines et al. 1992; Stefanova 1998b
Hippocampus, dentate gyrus	Madeira and Lieberman 1995
Striatum	Ovtscharoff et al. 1992; Bozhilova-Pastirova and Ovtscharoff 1996
Nucleus accumbens	Karamfiloff et al. 1998a,b
Anterior commissure	Allen and Gorski 1991
Corpus callosum	Mack et al. 1995
Diencephalon	
Hypothalamus Preoptic area	Gorski et al. 1978; Swaab and Hofman 1988; Davis et al. 1996
Interstitial nucleus of the anterior hypothalamus	LeVey 1991, 1993
Suprachiasmatic nucleus	Swaab et al. 1985
Supraoptic nucleus	Madeira and Lieberman 1995
Ventromedial nucleus	Madeira and Lieberman 1995
Arcuate nucleus	Madeira and Lieberman 1995
Midbrain	
Substantia nigra	Raab et al. 1995a,b
Pons	
Locus coeruleus	Babstock et al. 1997
Spinal cord	
Spinal nucleus of the bulbocavernous muscle	Breedlove and Arnold 1980; Forger and Breedlove 1986
Onuf nucleus	Forger et al. 1996

dimorphism of the zone controlling singing in zebra finch, which is five to six times larger in males than in females (Nottebohm and Arnold 1976). Konishi (1989) has established that the volume of this zone is directly dependent on gonadal steroids. Analogously, the frog vocal system is innervated by sexually dimorphic brain stem nuclei, where the number of motoneurons and axons to the larynx is greater in males and is under sex steroid control (Kelley and Dennison 1990). Another example for a direct connection between sex differences in structure and function is the spinal nucleus of the bulbocavernous muscle, which contains a greater number of neurons in male rats (Breedlove and Arnold 1980).

Many other sex differences are found in the brain of mammals, which are difficult to correlate with known functions because of the complicated circuitry underlying these behaviors. Most often these are sex differences in nuclear volume, numbers/densities of neurons, expression of neurotransmitters, receptors, and growth factors, magnitude of cell perikarya, neurites, and ultrastructural characteristics (organelles, synapses). Such differences are reported in various parts of the CNS (Table 18), and the present study adds new information on sex differences in the BST and in the amygdala of rats.

Generally, two patterns of sex differences are described in the mammalian brain. Either males have greater dimensions than females, or the opposite pattern exists – females show greater morphometric measurements than males. For example, the preoptic area is considered to play a key role in the control of reproductive behavior. Gorski et al. (1978) have found an outstanding gender difference in a zone of the preoptic area that they have named the sexually dimorphic nucleus of the preoptic area (SDN-POA). This nucleus has a five-to-six-times larger volume in male than in female rats and contains a greater number of neurons. Similar results have been obtained in other species including humans (Swaab and Hofman 1988). It is established that sex differences of the SDN-POA are dependent on testosterone presence in the critical period of sexual differentiation of the brain. We confirm in this study that the volume of the medial amygdala has a similar pattern of sexual differentiation. The opposite pattern of sexual dimorphism is observed in the anteroventral periventricular nucleus of the rat hypothalamus. Females show a greater number of neurons than males, and although it is sensitive to hormonal environment in the critical period of sexual differentiation, the difference is manifested around puberty. Similarly, the parastrial nucleus of the preoptic area has a greater volume in female than in male rats and it depends on gonadal steroids during early postnatal development (Del Abril et al. 1990).

Another group of sex differences concerns the neurotransmitter systems. Most often, such differences have been described in the hypothalamus and the limbic system, which participate in the control of reproductive behaviors. The anteroventral periventricular nucleus of the hypothalamus contains a greater number of cholecystokinin (CCK), tyrosine hydroxylase (TH), and corticotropin-releasing factor (CRF)-expressing neurons in females, while males have a greater number of enkephalin-expressing cells (Micevych et al. 1987; Simerly et al. 1985, 1988; McDonald et al. 1994). Oxytocin and prolactin-containing neurons are more numerous in the female preoptic area (Haussler et al. 1990; Beyer et al. 1992). It has been established that axon terminals, which contain GABA, neurotensin, CCK, methionine-enkephalin, galanin, TH, and dopamine, in the medial preoptic area are dependent on estrogen levels (Yuri and Kawata 1994a,b). Delville and Ferris (1995) have described a higher

Table 19. Some examples of sexual dimorphism of the human CNS

Some human CNS structures and/or functions with sex differences	References
Onuf nucleus	Forger and Breedlove 1986
Interstitial nucleus of the anterior hypothalamus	LeVay 1991, 1993
Cerebral cortex	Witelson et al. 1995; Pakkenberg and Gundersen 1997
Cerebral volume, caudate nucleus, globus pallidus, putamen, amygdala	Giedd et al. 1997
Anterior commissure	Allen and Gorski 1991, 1992
Splenium of corpus callosum	De Lacoste-Utamsing and Holloway 1982
Interhemispheric functional connections	Azari et al. 1995
Tourette's syndrome pathogenesis	Peterson et al. 1992

binding activity of arginine-vasopressin receptors in the ventrolateral hypothalamus of male hamsters. Our study extends the knowledge on sexual dimorphism of neurotransmitter systems with data on GABA, parvalbumin, and leucine-enkephalin expression in the BST and amygdala of the rat.

Sexual dimorphism has also been found in brain areas that are not occupied with the control of reproductive behavior. The female rat striatum has shown a greater number of GABA-immunoreactive neurons during fetal development and in adulthood (Ovtscharoff et al. 1992, 1997; Bozhilova-Pastirova and Ovtscharoff 1996). On the contrary, CCK-containing cells are more numerous in males (Pfister et al. 1989). In the male striatum and nucleus accumbens a higher overproduction and elimination of D1 and D2 dopamine receptors has been described (Andersen et al. 1997).

Sexual dimorphism is found not only in grey matter, but also in white matter formations. The corpus callosum shows a gender different midsagittal area (Berrebi et al. 1988) and different distribution of axonal types with prevalence of unmyelinated ones in females (Mack et al. 1995). The results of Lustig et al. (1991, 1993) suggest that GAP-43 mRNA and SNAP-25 mRNA could be used as markers for sexually specific synaptogenesis.

Gonadal steroids have regulatory effects on some growth factors. As a result, a higher concentration of the β-fraction of nerve growth factor has been established in the male cerebellum, olfactory bulb, hypothalamus, and the hypophysis compared to females (Ketoh-Semba et al. 1994). Recently, a modulatory effect of estrogen on Bcl-xL expression and apoptosis has been reported that might contribute to sexual differentiation as well as to age-related neurodegenerative conditions (Pike 1999)

Glial elements have also been shown to express sexual dimorphism. Garcia-Segura et al. (1988) described that the glial fibrillary acidic protein (GFAP) is under sex steroids control. The GFAP surface density in the hippocampus and globus pallidus is greater in male rats than in females. It is accepted that glia is one of the mediators of gonadal steroid action on the nervous system (see Jordan 1999).

Sexual dimorphism has also been described in the human CNS (Table 19). Some of these differences have been related to functional differences, sexual orientation, and pathology of the CNS, while others could not be sufficiently explained in terms of their functional significance.

5 Summary

The conclusions presented in this chapter summarize the original results of our study. They state the new data or data that support previous studies in the literature.

5.1
The Bed Nucleus of the Stria Terminalis

The BST is a small and yet heterogeneous limbic neuronal group which has shown to be sexually dimorphic in several aspects of its structure. The volume of the BST and the total neuronal density of its medial, lateral, and ventral subdivisions do not differ among genders. However, in the medial BST a subpopulation of neurons is present, which is sensible to androgen levels during the perinatal period. This statement is supported by our results on the decrease of the neuronal density in the medial BST of hormonally manipulated newborn male rats. The BST is rich in small and medium-sized GABA-immunoreactive neurons, whose density is greater in the female than in the male rat of different age groups. In the medial BST, androgens control the density of GABA-immunoreactive neurons both during the perinatal period through aromatization to estrogens as well as later after puberty. In the lateral and ventral BST, androgens exert only activational effects on the density of GABA-immunoreactive neurons late after puberty. The density of GABA-immunoreactive neurons increases with ageing, supposing a higher expression of GABA in the surviving neurons of the BST. Parvalbumin, which is generally considered to be often co-expressed by GABAergic neurons, is weakly expressed by neurons of the BST. A few parvalbumin-immunoreactive small neurons can be observed in the ventrolateral portion of the BST. They are most probably a small subpopulation of interneurons with weakly stained perikarya and their dendritic beginnings. The density of these parvalbumin-immunoreactive cells does not depend on androgen levels and shows no sex difference. The population of leucine-enkephalin-immunoreactive neurons comprises small and medium-sized cells. The density of leucine-enkephalin-immunoreactive neurons is sexually dimorphic only in the lateral BST early after birth (sixth day) and in later ages pre- and postpubertally, showing greater values in the male. The experimental manipulations of the hormonal environment strongly suggest that the density of leucine-enkephalin-immunoreactive neurons in the lateral BST is under the organizational effect of aromatizable androgens during the early perinatal period.

5.2
The Amygdala

Sexual dimorphism of four amygdaloid nuclei has been examined, namely the medial, cortical, basolateral, and central amygdala. The latter two are representatives of the deep amygdaloid nuclei, while the other two are representatives of the cortical structures connected with the amygdala. Only the medial amygdala has a sexually dimorphic volume. It is greater in males than in female rats as established in 20-day-, 3-month-, and 1-year- old animals. It is confirmed that sexual differentiation of the volume of the medial amygdala is controlled by aromatizable androgens during the critical perinatal period of life. Sex differences are observed also in the neuronal density of pre- and postpubertal medial and cortical amygdala. Male rats have a greater density of neurons in these amygdaloid nuclei than females. In the medial and cortical amygdala the sex difference is settled under the influence of androgens which are aromatized to estrogen during perinatal life. The population of small and medium-sized GABA-immunoreactive neurons in the rat amygdala shows sexually dimorphic density pre- and postpubertally in the medial and cortical amygdala, as well as in postpubertal central and basolateral amygdala, with females showing greater values than males. Diverse mechanisms controlling the expression of these sex differences are found. In the medial and cortical amygdala early postnatal organizational and late activational effects of androgens contribute to the formation of gender-specific density of GABA-immunoreactive neurons. In the basolateral amygdala the density of GABA-immunoreactive neurons is controlled by postpubertal levels of androgens, while in the central amygdala it is proposed that androgens have no direct influence on the density of GABA-immunoreactive neurons. Parvalbumin is expressed in a significant part of the small and medium-sized neurons in the medial, cortical, and basolateral amygdala. No parvalbumin-immunoreactive neurons are observed in the central amygdala. Sex difference in the density of parvalbumin-immunoreactive neurons is found in the cortical amygdala with higher values in males. It is suggested that androgen levels perinatally and postpubertally are responsible for the formation of the sexually dimorphic density of parvalbumin-immunoreactive neurons in the cortical amygdala. Leucine-enkephalin-immunoreactive neurons in the medial, cortical, central, and basolateral amygdala express sex difference only in postpubertal rats. Males show greater densities than females. The experimental manipulations of hormonal environment in male rats reveal complex mechanisms of androgen influence on the density of leucine-enkephalin-immunoreactive neurons. In the medial and central amygdala, predominantly late activational effects are responsible for the sexual differentiation of this neuronal subpopulation, while in the cortical and basolateral amygdala, late activational effects converge with early perinatal influences of aromatizable androgens.

Our results prove that the process of sexual differentiation of the brain is controlled by diverse mechanisms, most probably including not only hormonal but also other environmental and genomic factors. The data presented here try to explain the role of androgens during different stages of life span and demonstrate that circulating male gonadal hormones could be crucial for the sexual differentiation of various brain structures perinatally as well as in adulthood.

References

Alcantara S, Ferrer I, Soriano E (1993) Postnatal development of parvalbumin and calbidin D28 K immunoreactivities in the cerebral cortex of the rat. Anat Embryol Berl 188: 63–73

Allen L, Gorski RA (1990) Sex difference in the bed nucleus of the stria terminalis of the human brain. J Comp Neurol 302: 697–706

Allen L, Gorski RA (1991) Sexual dimorphism of the anterior commissure and massa intermedia of the human brain. J Comp Neurol 312: 97–104

Allen L, Gorski RA (1992) Sexual orientation and the size of the anterior commissure in the human brain. Proc Natl Acad Sci USA 89: 7199–7202

Allen LS, Hines M, Shryne JE, Gorski RA (1989) Two sexually dimorphic cell groups in the human brain. J Neurosci 9: 497–506

Al-Shamma HA, De Vries G (1996) Neurogenesis of the sexually dimorphic vasopressin cells of the bed nucleus of the stria terminalis and amygdala of rats. J Neurobiol 29: 91–98

Andersen SL, Rutstein M, Benzo JM, Hostetter JC, Teicher MH (1997) Sex differences in dopamine receptor overproduction and elimination. Neuroreport 8: 1495–1498

Andy OJ, Stephan H (1968) The septum in the human brain. J Comp Neurol 133: 383–410

Azari NP, Pettigrew KD, Pietrini P, Murphy DG, Horwitz B, Schapiro MB (1995) Sex differences in patterns of hemispheric cerebral metabolism: A multiple regression/discriminant analysis of positron emission tomographic data. Int J Neurosci 81: 1–20

Babstock D, Malsbury CW, Harley CW (1997) The dorsal locus coeruleus is larger in male than in female Sprague-Dawley rats. Neurosci Lett 224: 157–160

Baldino Jr F, McElligott SF, O'Kane TM, Gozes I (1988) Hormonal regulation of somatostatin messenger RNA. Synapse 2: 317–325

Barley J, Ginburg M, Greenstein BD, MacLusky NJ, Thomas P (1975) Androgen receptors in rat brain and pituitary. Brain Res 100: 383–393

Barraclough CA, Gorski RA (1962) Studies on mating behaviour in the androgen-sterilized female rat in relation to the hypothalamic regulation of sexual behaviour. J Endocrinol 25: 175–182

Beatty WW (1979) Gonadal hormones and sex differences in nonreproductive behaviors in rodents: Organizational and activational influences. Horm Behav 12: 112–163

Behzadi G, Kalen P, Parvopassu F, Wiklud L (1990) Afferents to the median raphe nucleus of the rat: retrograde cholera toxin and wheat germ conjugated horseradish peroxidase tracing, and selective D-[3H]aspartate labelling of possible excitatory amino acid inputs. Neurosci 37: 77–100

Ben-Ari Y, Zigmond RE, Moore KE (1975) Regional distribution of tyrosine hydroxylase, norepinephrine and dopamin within the amygdaloid complex of the rat. Brain Res 87: 96–101

Bergen HT, Hejtmancik JF, Pfaff DW (1991) Effects of gamma-aminobutyric acid receptor agonist and antagonist on LHRH-synthesizing neurons as detected by immunohistochemistry and in situ hybridization. Exp Brain Res 87: 46–56

Berk ML, Finkelstein JA (1981) Afferent projections to the preoptic area and hypothalamic regions in the rat brain. Neuroscience 6: 1601–1624

Berrebi AS, Fitch RH, Ralphe DL, Denenberg JO, Friedrich Jr VL, Denenberg VH (1988) Corpus callosum: Region-specific effects of sex, early experience and age. Brain Res 438: 216–224

Beyer C, Hutchison JB (1997) Developmental profile and regulation of brain estrogen synthesis by aromatase. Biomed Rev 7: 41–50

Beyer C, Kolbinger W, Froehlich U, Pilgrim C, Reisert I (1992) Sex differences of hypothalamic prolactin cells develop independently of the presence of sex steroids. Brain Res 593: 253–256

Beyer C, Wozniak A, Hutchison JB (1993) Sex-specific aromatization of testosterone in mouse hypothalamic neurons. Neuroendocrinol 58: 673–681

Björklund A, Lindvall O (1984) Dopamine-containing systems in the CNS. In: Björklund A, Hökfelt T, (eds.) Handbook of chemical neuroanatomy, vol. 2: Classical transmitters in the CNS, Part I. Amsterdam, Elsevier, pp. 55–122

Bozhilova-Pastirova A, Ovtscharoff W, (1996) Sex differences among GABA-immunoreactive and parvalbumin-immunoreactive neurons of the rat striatum. 91: Versammlung in Jena, Verh. Anat. Ges., Ann. Anat. 178: Suppl., p. 203

Bozhilova-Pastirova A, Ovtscharoff W, Brasizova D, Stefanova N (1997a) Age-related changes of morphological characteristics among neurons in the rat nucleus accumbens. Ann Anat 179: 180

Bozhilova-Pastirova A, Ovtscharoff W, D. Brazitsova, Stefanova N (1997b) Immunohistochemical study of the sex differences in the development of rat sensorimotor cortex. XIII Congress of the Bulgarian Anatomical Society, Varna, Bulgaria, Eur. J. Morph. 37: p. 56

Braak H, Braak E (1983) Neuronal types in the basolateral amygdaloid nuclei of man. Brain Res Bull 11: 349–365

Breedlove SM, Arnold AP (1980) Hormone accumulation in a sexually dimorphic motor nucleus in the rat spinal cord. Science 210: 564–566

Breedlove SM, Cooke BM, Jordan CL (1999) The orthodox view of brain sexual differentiation. Brain Behav Evol 54: 8–14

Bressler SC, Baum MJ (1996) Sex comparison of neuronal Fos immunoreactivity in the rat vomeronasal projection circuit after chemosensory stimulation. Neuroscience 71: 1063–1072

Brockhaus H. (1938) Zur normalen und patologischen Anatomie des Mandelkerngebietes. J. Psychol. Neurol. 49: 1–136

Brodal P (1998) The amygdala and emotions. In: The central nervous system: structure and function, Oxford, Oxford University Press, pp.559–565

Burroughs LF, Fiber JM, Swann JM (1996) Neuropeptide Y in hamster limbic nuclei: Lack of colocalization with substance P. Peptides 17: 1053–1062

Caffe AR, Van Leeuwen FW (1983) Vasopressin-immunoreactive cells in the dorsomedial hypothalamic region, medial amygdaloid nucleus and locus coeruleus of the rat. Cell Tiss Res 233: 23–33

Caffe AR, van Leeuwen FW, Luiten PG (1987) Vasopressin cells in the medial amygdala of the rat project to the lateral septum and ventral hippocampus. J Comp Neurol 261: 237–252

Canteras NS, Simerly RB, Swanson LW (1992) Projections of the ventral premammillary nucleus. J Comp Neurol 324: 195–212

Canteras NS, Simerly RB, Swanson LW (1995) Organization of projections from the medial nucleus of the amygdala: a PHAL study in the rat. J Comp Neurol 360: 213–245

Canteras NS, Swanson LW (1992) Projections of the ventral subiculum to the amygdala, septum, and hypothalamus: a PHAL anterograde tract-tracing study in the rat. J Comp Neurol 324: 180–194

Cassell MD, Gray TS (1989) Morphology of peptide-immunoreactive neurons in the rat central nucleus of the amygdala. J Comp Neurol 281: 320–333

Celio MP (1990) Calbindin-D28 k and parvalbumin in the rat nervous system. Neurosci 35: 375–475

Choate JV, Resko JA (1992) Androgen receptor immunoreactivity in intact and castrated guinea pig using antipeptide antibodies. Brain Res 597: 51–59

Churchill L, Zahm DS, Kalivas PW (1996)The mediodorsal nucleus of the thalamus in rats–I. forebrain GABAergic innervation. Neuroscience 70: 93–102

Clancy AN, Bonsall RW, Michael RP (1992) immunohistochemical labelling of androgen receptors in the brain of rat and monkey. Life Sci 50: 409–417

Clancy AN, Whitman C, Michael RP, Albers HE (1994) Distribution of androgen receptor like-immunoreactivity in the brains of intact and castrated male hamsters. Brain Res Bull 33:325–332

Conde F, Lund JS, Jacobowitz DM, Baimbridge KG, Lewis DA (1994) Local circuit neurons immunoreactive for calretin, calbidin D-28 k or parvalbumin in monkey prefrontal cortex: distribution and morphology. J Comp Neurol 341: 95–116

Cooke BM, Tabibnia G, Breedlove SM (1999) A brain sexual dimorphism controlled by adult circulating androgens. Proc Natl Acad Sci USA 96: 7538–7540

Cowan RL, Wilson CJ, Emson PC, Heizmann CW (1990) Parvalbumin-containing GABAergic interneurons in the rat neostriatum. J Comp Neurol 302: 197–205

Cullinan WE, Herman JP, Watson SJ (1993) Ventral subicular interaction with the hypothalamic paraventricular nucleus: evidence for a relay in the bed nucleus of the stria terminalis. J Comp Neurol 332: 1–20

Da Costa Gomez T, Behbehani M (1995) An electrophysiological characterization of the projection from the central nucleus of the amygdala to the periaqueductal gray of the rat: the role of opioid receptors. Brain Res 689: 21–31

Davis EC, Shryne JE, Gorski RA (1996) Structural sexual dimorphisms in the anteroventral periventricular nucleus of the rat hypothalamus are sensitive to gonadal steroids perinatally, but develop peripubertally. Neuroendocrinol 63: 142–148

De Lacoste-Utamsing C, Holloway RL (1982) Sexual dimorphism in the human corpus callosum. Science 216: 1431–1432

De Olmos J, Alheid GF, Beltramino CA (1985) Amygdala. In G. Paxinos (ed): The rat nervous system, Vol. 1.Forebrain and midbrain. Orlando:Academic Press, pp.223–334

De Olmos J. (1990) Amygdaloid nuclear grey complex. In: Paxinos G. (ed.): The human nervous system. Academic Press, San Diego, 583–710

de Quidt ME, Kiyama H, Emson PC (1990) Pancreatic polypeptide, neuropeptide Y and peptide YY in central neurons. In: Bjorklund A, Hokfelt T, Kuhar MJ (eds.) Handbook of chemical neuroanatomy, vol. 9: Neuropeptides in the CNS, Part II. Amsterdam, Elsevier, pp. 287–358

De Vries GJ (1990) Sex differences in neurotransmitter systems. J Neuroendocrinol 2: 1–13

De Vries GJ, Al-Shamma HA (1990) Sex differences in hormonal responses of vasopressin pathways in the rat brain. J Neurobiol 21: 686–693

Del Abril A, Segovia S, Guillamón (1987) The bed nucleus of the stria terminalis in the rat; regional sex differences controlled by gonadal steroids early after birth. Dev Brain Res 32: 295–300

Del Abril A, Segovia S, Guillamón A (1990) Sexual dimorphism in the parastrial nucleus of the rat preoptic area. Dev Brain Res 52: 11–15

Del Cerro MCR, Izquierdo MAP, Perez-Laso C, Rodriguez-Zafra M, Guillamon A, Segovia S (1995) Early postnatal diazepam exposure facilitates maternal behavior in virgin female rats. Brain Res Bull 38: 143–148

Delville Y, Ferris CF (1995) Sexual differences in vasopressin receptor binding within the ventrolateral hypothalamus in golden hamsters. Brain Res 681: 91–96

Demling J, Fuchs Baumert M, Wuttke W (1985) Preoptic catecholamine, GABA, and glutamate release in ovariectomized and ovariectomized estrogen-primed rats utilizing a push-pull cannula technique. Neuroendocrinol 41: 212–218

Don Carlos LL, McAbee M, Ramer-Quinn DS, Stancik DM (1995) Estrogen receptor mRNA levels in the preoptic area of neonatal rats are responsive to hormone manipulation. Dev Brain Res 84: 253–260

Fallon JH, Moore RY (1978) Catecholamine innervation of the basal forebrain. IV. Topography of the dopamine projection of the basal forebrain and neostriatum. J. Comp. Neurol. 180: 545–580

Flügge G, Oertel WH, Wuttke W (1986) Evidence for estrogen receptive GABAergic neurons in the preoptic/anterior hypothalamic area of the rat brain. Neuroendocrinol 43: 1–5

Forger NG, Breedlove SM (1986) Sexual dimorphism in human and canine spinal cord: Role of early androgen. Proc. Natl Acad Sci USA 83: 7527–7531

Forger NG, Frank LG, Breedlove SM, Glickman SE (1996) Sexual dimorphism of perineal muscles and motoneurons in spotted hyenas. J Comp Neurol 375: 333–343

Fortin G, Jungbluth S, Lumsden A, Champagnat J (1999) Segmental specification of GABAergic inhibition during development of hindbrain neural networks. Nat Neurosci 2:873–877

Fox CA (1949) Amygdalo-thalamic connections in Macaca mulatta. Anat. Rec. 103: 537–538

Frankfurt M, Fuchs E, Wuttke W (1984)Sex differences in gamma-aminobutyric acid and glutamate concentrations in discrete rat brain nuclei. Neurosci Lett 50:245–250

Freund TF (1989) GABAergic septohippocampal neurons contain parvalbumin. Brain Res 478: 375–381

Fuller TA, Russchen FT, Price JL (1987) Source of presumptive glutamatergic/aspartatergic afferents to the rat ventral striatopallidal region. J Comp Neurol 258: 317–338

Garcia-Segura LM, Suarez I, Segovia S, Tranque PA, Cales JM, Aguilera P, Olmos G, Guillamon A (1988) The distribution of glial fibrillary acidic protein in the adult rat brain is influenced by the neonatal levels of sex-steroids. Brain Res 456: 357–363

Giedd JN, Castellanos FX, Rajapakse JC, Vaituzis AC, Rapoport JL (1997) Sexual dimorphism of the developing human brain. Prog Neuro-Psychopharmacol Biol Psychiat 21: 1185–1201

Glezer II, Hof PR, Leranth C, Morgane PJ (1993) Calcium-binding protein-containing neuronal populations in mammalian visual cortex: a comparative study in whales, insectivores, bats, rodents, and primates. Cereb Cortex 3: 249–272

Gloor P. (1960) Amygdala. In: Field J, Magoun HW, Hall V, Handbook of physiology. Section I: Neurophysiology, American Physiological Society, Washington, 1395–1420

Gorski RA (1985 a) Sexual differentiation of the brain: possible mechanisms and implications. Can J Physiol Pharmacol 63: 577–594

Gorski RA (1985 b) Sexual dimorphisms of the brain. J. Anim Sci 61:38–61

Gorski RA, Gordon JH, Shryne JE, Southam AM (1978) Evidence for a morphological sex difference within the medial preoptic area of the rat brain. Brain Res 148: 333–346

Gorski RA, Harla RE, Jacobson CD, Shryne JE, Southam AM (1980) Evidence for the existence of a sexually dimorphic nucleus in the preoptic region of the rat. J Comp Neurol 193: 529–539

Gould E, Tanapat P, McEwen BS, Flügge G, Fuchs E (1998) Proliferation of granule cell precursors in the dentate gyrus of adult monkeys is diminished by stress. Proc Natl Acd Sci USA 95: 3168–3171

Grattan DR, Selmanoff M (1993) Regional variation in γ-aminobutyric acid turnover: Effect of castration on γ-aminobutyric acid turnover in microdissected brain regions of the male rat. J Neurochem 60: 2254–2264

Grattan DR, Selmanoff M (1994 a) Castration-induced decrease in the activity of medial preoptic and tuberoinfundibular GABAergic neurons is prevented by testosterone. Neuroendocrinol 60: 141–149

Grattan DR, Selmanoff M (1994 b) Prolactine- and testosterone-induced inhibition of LH secretion after orchidectomy: role of preoptic and tuberoinfundibular γ-aminobutyric acidergic neurones. J Endocrinol 143: 165–174

Gray TS, Magnuson DJ (1987) Neuropeptide neuronal efferents from the bed nucleus of the stria terminalis and central amygdaloid nucleus to the dorsal vagal complex in the rat. J Comp Neurol 262: 365–374

Gritti I, Mainville L, Jones BE (1994) Projections of GABAergic and cholinergic basal forebrain and GABAergic preoptic-anterior hypothalamic neurons to the posterior lateral hypothalamus of the rat. J Comp Neurol 339: 251–268

Gritti I, Mariotti M, Mancia M (1998) GABAergic and cholinergic basal forebrain and preoptic-anterior hypothalamic projections to the mediodorsal nucleus of the thalamus in the cat. Neuroscience 85: 149–178

Guillamón A, Segovia S, del Abril A (1988) Early effects of gonadal steroids on the neuron number in the medial posterior region and the lateral division of the bed nucleus of the stria terminalis in the rat. Dev Brain Res 44: 281–290

Hall E (1972) Some aspects of the structural organization of the amygdala. In: Eleftheriou BE (ed.) The neurobiology of the amygdala. Plenum press, New York, pp. 95–120

Hansen S, Ferriera A, Selart ME (1985) Behavioral similarities between mother rats and benzodiazepine treated non-maternal animals. Psychopharmacol 86: 344–347

Hashimoto R, Kimura F (1986) Inhibition of gonadotropin secretion induced by cholecystokinin implants in the medial preoptic area by the dopamine receptor blocker, pimozide, in the rat. Neuroendocrinol 42: 32–37

Haussler HU, Jirikowski GF, Caldwell JD (1990) Sex differences among oxytocin immunoreactive neuronal systems in the mouse hypothalamus. J Chem Neuroanat 3: 271–276

Heizmann CW (1984) Parvalbumin, an intracellular calcium-binding protein; distribution, properties and possible roles in mammalian cells. Experientia 40: 910–921

Herbison AE, Chapman C, Dyer RG (1991 a) Role of medial preoptic GABA neurons in regulating luteinizing hormone secretion in the ovariectomized rat. Exp Brain Res 87:345–352

Herbison AE, Heavens RP, Dye S, Dyer RG (1991 b) Acute action of oestrogen on medial preoptic gamma-aminobutyric acid neurons: Correlation with oestrogen negative feedback on luteinizing hormone secretion. J Neuroendocrinol 3: 101–106

Herman JP, Cullinan WE (1997) Neurocircuitry of stress: central control of the hypothalamo-pituitary-adrenocortical axis. Trends Neurosci 20: 78–84

Hines M, Allen LS, Gorski RA (1992) Sex differences in subregions of the medial nucleus of the amygdala and the bed nucleus of the stria terminalis of the rat. Brain Res 579: 321–326

Hines M, Davis FC, Coquelin A, Goy RW, Gorski RA (1985) Sexually dimorphic regions in the medial preoptic area and the bed nucleus of the stria terminalis of the guinea pig brain: a description and an investigation of their relationship to gonadal steroids in adulthood. J Neurosci 5: 40–47

Holstege G, Meiners L, Tan K (1985) Projections of the bed nucleus of the stria terminalis to the mesencephalon, pons and medulla oblongata in the cat. Exp Brain Res 58: 379–391

Hopkins DA, Holstege G (1978) Amygdaloid projections to the mesencephalon, pons and medulla oblongata in the cat. Exp. Brain Res 32: 529–547

Horvath TL, Naftolin F, Leranth C (1993) Luteinizing hormone –releasing hormone and gamma-aminobutyric acid neurons in the medial preoptic area are synaptic targets of dopamine axons originating in the anterior periventricular area. J Neuroendocrinol 5: 71–79

Hutchison JB, Beyer C, Hutchison RE, Wozniak A (1995) Sexual dimorphism in the developmental regulation of brain aromatase. J Steroid Biochem Mol Biol 53: 307–313

Hutton LA, Gu G, Simerly RB (1998) Development of a sexually dimorphic projection from the bed nuclei of the stria terminalis to the anteroventral periventricular nucleus in the rat. J Neurosci 18: 3003–3013

Iqbal J, Swanson JJ, Prins GS, Jacobson CD (1995) Androgen-receptor-like immunoreactivity in the Brazilian opossum brain and pituitary: distribution and effects of castration and testosterone replacement in the adult male. Brain Res 703: 1–18

Jakab RL, Horvath TL, Leranth C, Harada N, Naftolin F (1993) Aromatase immunoreactivity in the rat brain: Gonadectomy-sensitive hypothalamic neurons and an unresponsive "limbic ring" of the lateral septum-bed nucleus-amygdala complex. J Steroid Biochem Molec Biol 44: 481–498

Johnston JB (1923) Further contribution to the study of the evolution of the forebrain. J Comp Neurol 35: 337–481

Jolkkonen E, Pitkänen A (1998) Intrinsic connections of the rat amygdaloid complex: Projections originating in the central nucleus. J Comp Neurol 395: 53–72

Jordan CL (1999) Glia as mediators of steroid hormone action on the nervous system: An overview. J Neurobiol 40: 434–435

Ju G, Swanson LW (1989) Studies on the cellular architecture of the bed nuclei of the stria terminalis in the rat: I. Cytoarchitecture. J Comp Neurol 280: 587–602

Ju G, Swanson LW, Simerly RB (1989) Studies on the cellular architecture of the bed nucleus of the stria terminalis in the rat: II. Chemoarchitecture. J Comp Neurol 280: 603–621

Jüptner M, Hiemke C (1990) Sex differences of GABA$_A$ receptor binding in rat brain measured by an improved in vitro binding assay. Exp Brain Res 81:297–302

Kalimullina LB (1985) Reaction of the neurons of the basolateral nuclear group of the amygdaloid complex to gonadectomy in rat. Arch Anat Histol Embryol, LΓVIII, 3: 20–25

Kalimullina LB (1986) Regions of sexual dimorphism in the cortico-medial group of the amygdaloid complex. Arch Anat Histol Embryol, XCI, 9: 22–26

Kalimullina LB (1988) Regions of sexual dimorphism in the basolateral structures of the amygdaloid complex. Arch Anat Histol Embryol, XCV, 8: 17–21

Karamfiloff K, Baeva N, Stoykov I, Kolev J, Stefanova N, Bozhilova-Pastirova A, Ovtscharoff W(1998) Sex differentiation of parvalbumin-immunoreactive neurons in the rat nucleus accumbens. Symposium of clinical anatomy, Varna, Scripta Sci. Med., Γ, Suppl., p. 101

Katoh-Semba R, Semba R, Kato H, Ueno M, Arakawa Y, Kato K (1994) Regulation by androgen of levels of the β subunit of nerve growth factor and its mRNA in selected regions of the mouse brain. J Neurochem 62: 2141–2147

Kawata M (1995) Roles of steroid hormones and their receptors in structural organization in the nervous system. Neurosci Res 24: 1–46

Kelce MJ, Genjam VK, Rudeen PK (1980) Effects of fetal alcohol exposure on brain 5α-reductase/aromatase activity. J Steroid Biochem 35:103–106

Kelley DB, Dennison J (1990) The vocal motor neurons of *Xenopus laevis*: development of sex differences in axon number. J Neurobiol 21: 869–882

Kelly DD (1991) Sexual differentiation of the nervous system. In: Kandel ER, Schwartz JH (ed.) Principles of neural science. 960–873

Kimura F, Hashimoto R, Kawakami M (1983) The stimulatory effect of cholecystokinin implanted in the medial preoptic area on luteinizing hormone secretion in the ovariectomized estrogen-primed rat. Endocrinol Jpn 30: 305–309

Koch M, Ehret G (1989) Immunocytochemical localization and quantitation of estrogen-binding cells in the male and female (virgin, pregnant, lactating) mouse brain. Brain Res 489: 101–112

Kolev J, I Stoykov, Stefanova N, Bozhilova-Pastirova A,, Ovtscharoff W. (1997) Parvalbumin-immunoreactive neurons in the rat striatum: Light and Electron microscopic study. Ann. Anat., 179: Suppl., p.198–199

Kondo Y, Yamanouchi K (1995) The possible involvement of the nonstrial pathway of the amygdala in neural control of sexual behavior in male rats. Brain Res Bull 38: 37–40

Konishi M (1989) Birdsong for neurobiologists. Neuron 3: 541–549

Kostarczyk EM (1986) The amygdala and male reproductive functions: I. Anatomical and endocrine bases. Neurosci Biobehav Rev 10: 67–77

Krettek JE, Price JL (1978 a) A description of the amygdaloid complex in the rat and cat with observations on intra-amygdaloid axonal connections. J Comp Neurol 178: 255–280

Krettek JE, Price JL (1978 b) Amygdaloid projections to subcortical structures within the basal forebrain and brainstem in the rat and cat. J Comp Neurol 178: 225–254

Kühnemann S, Brown TJ, Hochberg RB, MacLusky NJ (1995) Sexual differentiation of estrogen receptor concentrations in the rat brain; effects of neonatal testosterone exposure. Brain Res 691: 229–234

Kuiper GGJM, Enmark E, Pelto-Huikko M, Nilsson S, Gustafsson J-A (1996) Cloning of a novel estrogen receptor expressed in rat prostate and ovary. Proc Natl Acad Sci USA 93: 5925–5930

Lahr G, Maxson SC, Mayer A, Just W, Pilgrim C, Reisert I (1995) Transcription of the Y chromosomal gene, Sry, in adult mouse brain. Mol Brain Res 33: 179–182

Lakhdar-Ghazal N, Dubois-Dauphin M, Hermes M, Buijs R, Bengelloun W, Pevet P (1995) Vasopressin in the brain of a desert hibernator, the jerboa (*Jaculus orientalis*): presence of sexual dimorphism and seasonal variation. J Comp Neurol 358: 499–517

Lasaga M, Duvilanski BH, Seilicovich A, Afione S, Debeljuk L (1988) Effect of sex steroids on GABA receptors in the rat hypothalamus and anterior pituitary gland. Eur J Pharmacol 155:163–166

Lephart ED (1996) A review of brain aromatase cytochrome P450. Brain Res Rev 22: 1–26

Lephart ED, Simpson ER, Mc Phaul MJ, Kilgore MW, Wilson JD, Ojeda SR (1992 a) Brain aromatase cytochrome P-450 messenger RNA and enzyme activity during prenatal and perinatal development in the rat. Mol Brain Res 16: 187–192

Lephart ED, Simpson ER, Ojeda SR (1992 b) Effects of cyclic AMP and androgens on in vitro brain aromatase enzyme activity during prenatal development in the rat. J Neuroendocrinol 4: 29–36

Letinic K, Kostovic I (1996) Transient neuronal population of the internal capsule in the developing human cerebrum. Neuroreport 7: 2159–2162

LeVay S (1991) A difference in hypothalamic structure between heterosexual and homosexual men. Science 253: 1934–1037

LeVay S (1993) The sexual brain. MIT Press

Lind RW, Ganten D (1990) Angiotensin. In: Bjorklund A, Hokfelt T, Kuhar MJ (eds.) Handbook of chemical neuroanatomy, vol. 9: Neuropeptides in the CNS, Part II. Amsterdam, Elsevier, pp 165–286

Lolova I, Davidoff M (1991) Changes in GABA-immunoreactivity and GABA-transaminase activity in rat amygdaloid complex in aging. J Hirnforsch 32: 231–238

Loughlin SE, Fallon JH (1983) Dopaminergic and non-dopaminergic projections to amygdala from substantia nigra and ventral tegmental area. Brain Res 262: 334–338

Loughlin SE, Fallon JH (1985) Locus coeruleus. In: Paxinos G (ed.) The rat nervous system. Academic Press, Australia, pp. 79–93

Lustig RH, Hua P, Wilson MC, Federoff HJ (1993) Ontogeny, sex dimorphism, and neonatal sex hormone determination of synapse-associated messenger RNAs in rat brain. Mol Brain Res 20: 101–110

Lustig RH, Sudol M, Pfaff DW, Federoff HJ (1991) Estrogenic regulation and sex dimorphism of growth-associated protein 43 kDa (GAP-43) messenger RNA in the rat. Mol Brain Res 11: 125–132

Mack CM, Boehm GW, Berrebi AS, Denenberg VH (1995) Sex differences in the distribution of axon types within the genu of the rat corpus callosum. Brain Res 697: 152–156

Madeira MD, Lieberman AR (1995) Sexual dimorphism in the mammalian nervous system. Prog Neurobiol 45: 275–333

Magnusson A, Dahlfors G, Blomqvist A (1996) Differential distribution of calcium-binding proteins in the dorsal column nuclei of rats. A combined immunohistochemical and retrograde tract tracing study. Neurosci 73: 497–508

Majewska DM, Ford-Rice F, Falkay G (1989) Pregnancy-induced alterations of GABAa receptor sensitivity in maternal brain: an antecedent of postpartum "blues"? Brain Res 482: 397–401

Makino S, Gold PW, Schulkin J (1994) Effects of corticosterone on CRH mRNA and content in the bed nucleus of the stria terminalis; comparison with the effects in the central nucleus of the amygdala and the paraventricular nucleus of the hypothalamus. Brain Res 657: 141–149

Malsbury CW, McKay K (1987) A sex difference in the pattern of substance P-like immunoreactivity in the bed nucleus of the stria terminalis. Brain Res 420: 365–370

Malsbury CW, McKay K (1989) Sex difference in the substance P-immunoreactive innervation of the medial nucleus of the amygdala. Brain Res Bull 23: 561–567

Martinez-Guijarro FG, Blasco-Ibàñez JM, Lopez-Garcia C (1994) Postnatal increase of GABA- and parvalbumin-immunoreactive cells in the cerebral cortex of the lizard *Podarcis hispanica*. Brain Res 634: 168–172

Maxson S (1997) Sex differences in genetic mechanisms for mammalian brain and behavior. Biomed Rev 7: 85–90

Mayer A, Lahr G, Swaab DF, Pilgrim C, Reisert I (1998) The Y-chromosomal genes SRY and ZFY are transcribed in adult human brain. Neurogenetics 1:281–288

McCarthy MM, Kaufman LC, Brooks PJ, Pfaff DW, Schwartz-Giblin S (1995) Estrogen modulation of mRNA levels for the two forms of glutamic acid decarboxylase (GAD) in female rat brain. J Comp Neurol 360: 685–697

McCarthy MM, Masters DB, Fiber JM, Lopez-Colome AM, Beyer C, Komisaruk BR, Feder HH (1991) GABAergic control of receptivity in the female rat. Neuroendocrinol. 533: 473–479

McDonald AJ (1982 a) Cytoarchitecture of the central amygdaloid nucleus of the rat. J Comp Neurol 208: 401–418

McDonald AJ (1982 b) Neurons of the lateral and basolateral amygdaloid nuclei: a Golgi study in the rat. J Comp Neurol 212: 295–312

McDonald AJ (1983) Neurons of the bed nucleus of the stria terminalis: A Golgi study in the rat. Brain Res Bull 10: 111–120

McDonald AJ (1984) Neuronal organization of the lateral and basolateral amygdaloid nuclei in the rat. J Comp Neurol 222: 589–606

McDonald AJ (1989) Coexistence of somatostatin with neuropeptide Y, but not with cholecystokinin or vasoactive intestinal peptide, in neurons of the rat amygdala. Brain Res 500: 37–45

McDonald AJ, Mascagni F (1997) Projections of the lateral entorhinal cortex to the amygdala: a Phaseolus vulgaris leucoagglutinin study in the rat. Neuroscience 77: 445–459

McDonald AJ, Mascagni F, Wilson MA (1994) A sexually dimorphic population of CRF neurons in the medial preoptic area. Neuroreport 5: 653–656

McDonald AJ, Pearson JC (1989) Coexistence of GABA and peptide immunoreactivity in non-pyramidal neurons of the basolateral amygdala. Neurosci Lett 100:53–58

McEwen BS (1991) Non-genomic and genomic effects of steroids on neural activity. TIPS 12: 141–147

McEwen BS, Lieberburg I, Chaptal C, Krey LC (1977) Aromatization: important for sexual differentiation of the neonatal rat brain. Horm Behav 9: 249–263

McLean S, Skirboll LR, Pert CB (1983) Opiatergic projection from the bed nucleus to the habenula: demonstration by a novel radioimmunohistochemical method. Brain Res 278: 255–257

Mehler WR (1980) Subcortical afferent connections of the amygdala in the monkey. J Comp Neurol 190: 733–762

Mendelson MD, Gorzalka BB (1984) Cholecystokinin-octapeptide produces inhibition of lordosis in the female rat. Pharmacol Biochem Behav 21: 755–759

Micevych P, Akesson T, Elde R (1988) Distribution of cholecystokinin-immunoreactive cell bodies in the male and female rat: II. Bed nucleus of the stria terminalis and amygdala. J Comp Neurol 269: 381–391

Micevych P, Matt DW, Go VL (1988) Concentrations of cholecystokinin, substance P, and bombesin in discrete regions of male and female rat brain: sex differences and estrogen effects. Exp Neurol 100:416–425

Micevych P, Park SS, Akesson TR, Elde R (1987) The distribution of cholecystokinin immunoreactive cell bodies in male and female rat: Hypothalamus. J Comp Neurol 269: 124–136

Micheva KD, Beaulieu C (1995) Postnatal development of GABA neurons in the rat sensorimotor barrel cortex: a quantitative study. Eur J Neurosci 7:419–430

Miller MA, Kolb PE, Raskind MA (1993) Testosterone regulates galanin gene expression in the bed nucleus of the stria terminalis. Brain Res 611: 338–341

Mizukami S, Nishizuka M, Arai Y (1983) Sexual difference in nuclear volume and ontogeny in the rat amygdala. Exp Neurol 79: 569–575

Mizuno N, Takahashi O, Satoda T, Matsushima R (1985) Amygdalospinal projections in the macaque monkey. Neurosci Lett 53: 327–330

Moga MM, Saper CB, Gray TS (1989) Bed nucleus of the stria terminalis; cytoarchitecture, immunohistochemistry, and projections to the parabrachial nucleus in the rat. J Comp Neurol 283: 315–332

Morin LP, Goodless-Sanchez N, Smale L, Moore RY (1994) Projections of the suprachiasmatic nuclei, subparaventricular zone and retrochiasmatic area in the golden hamster. Neurosci 61: 391–410

Mugnaini E, Oertel WH (1985) Atlas of the distribution of GABAergic neurons and terminals in the rat CNS. In: Bjorklund A, Hokfelt T (eds.). GABA and neuropeptides in the CNS, part I. Elsevier, Amsterdam, 1985; pp. 436–608

Naftolin F, Ryan KJ, Davies IJ, Reddy VV, Flores F, Petro Z, Kuhn M, White RJ, Takaoka Y, Wolin L (1975) The formation of estrogens by central neuroendocrine tissues. Rec Prog Horm Res 31: 795–796

Nauta WJH (1961) Fibre degeneration following lesions of the amygdaloid complex in the monkey. J Anat 95: 515–531

Nauta WJH (1962) Neural associations of the amygdaloid complex in the monkey. Brain 276: 237–245

Neal Jr. CR, Swann JM, Newman SW (1989) The colocalization of substance P and prodynorphin immunoreactivity in neurons of the medial preoptic area, bed nucleus of the stria terminalis and medial nucleus of the amygdala of the Syrian hamster. Brain Res 496: 1–13

Nishizuka M, Arai Y (1981) Sexual dimorphism in synaptic organization in the amygdala and its dependence on neonatal hormone environment. Brain Res 212: 31–38

Nitecka L, Ben-Ari Y (1987) Distribution of GABA-like immunoreactivity in the rat amygdaloid complex. J Comp Neurol 266: 45–55

Nottebohm F, Arnold AP (1976) Sexual dimorphism in vocal control areas of the songbird brain. Science 194: 211–213

Numan M, Numan MJ (1997) Projection of medial preoptic area and ventral bed nucleus of the stria terminalis neurons that express Fos during maternal behavior in female rats. J Neuroendocrinol 9: 369–384

O'Keefe JA, Handa RJ (1990) Transient elevation of estrogen receptors in the neonatal rat hippocampus. Dev Brain Res 57: 119–127

Ostrowski NL, Hill JM, Pert CB, Pert A (1987) Autoradiographic visualization of sex differences in the pattern and density of opiate receptors in hamster hypothalamus. Brain Res 421: 1–13

Otake K, Nakamura Y (1995) Sites of origin of corticotropin-releasing factor-like immunoreactive projection fibers to the paraventricular thalamic nucleus in the rat. Neurosci Lett 201: 84–86

Otake K, Ruggiero DA, Nakamura Y (1995) Adrenergic innervation of forebrain neurons that project to the paraventricular thalamic nucleus in the rat Brain Res 697: 17–26

Ottersen OP (1981) Afferent connections to the amygdaloid complex of the rat with some observations in the cat. III. Afferents from the lower brain stem. J Comp Neurol 202: 335–356

Ottersen OP (1982) Connections of the amygdala of the rat. IV: Corticoamygdaloid and intraamygdaloid connections as studied with axonal transport of horseradish peroxidase. J Comp Neurol 205: 30–48

Ovtscharoff W, Bozhilova-Pastirova A, Vankova M (1997) Sex differences in the development of the rat striatum. Eur. J Morphol. 35: p. 62

Ovtscharoff W, Eusterschulte B, Zienecker R, Reisert I, Pilgrim C (1992) Sex differences in densities of dopaminergic fibers and GABAergic neurons in the prenatal rat striatum. J Comp Neurol 323: 299–304

Pakkenberg B, Gundersen HJ (1997) Neocortical neuron number in humans: effect of sex and age. J. Comp. Neurol. 384: 312–320

Parducz A, Perez J, Garcia-Segura LM (1993) Estradiol induces plasticity of GABAergic synapses in the hypothalamus. Neurosci 53: 395–401

Paxinos G, Watson C (1986) The rat brain in stereotaxic coordinates. New York: Academic Press

Peterson BS, Leckman JF, Scahill L, Naftolin F, Keefe D, Charest N, Cohen D (1992) Steroid hormones and CNS sexual dimorphisms modulate symptom expression in Tourette's syndrome. Psychoneuroendocrinol 17: 553–563

Petit JM, Luppi PH, Peyron C, Rampon C, Jouvet M (1995) VIP-like immunoreactive projection from the dorsal raphe and caudal linear raphe nuclei to the bed nucleus of the stria terminalis demonstrated by a double immunohistochemical method in the rat. Neurosci Lett 193: 77–80

Petrovich GD, Risold PY, Swanson LW (1996) Organization of projections from the basomedial nucleus of the amygdala: a PHAL study in the rat. J Comp Neurol 374: 387–420

Pfaff D, Keiner M (1973) Atlas of estradiol-concentrating cells in the central nervous system of the female rat. J Comp Neurol 151: 121–158

Pfeiffer CA (1936) Sexual differences of the hypophyses and their determination by the gonads. Am J Anat 58: 195–226

Pfister C, Schade R, Ott T (1989) Sexually dimorphic level of CCK-8-like immunoreactive neuronal somata within several basal forebrain nuclei of the rat. Exp Clin Endocrinol 94: 121–126

Phoenix CH, Giy RW, Gerall AA, Young WC (1959) Organizing action of prenatally administered testosterone propionate on the tissues mediating mating behavior in the female guinea pig. Endocrinology 65: 369–382

Pike CJ (1999) Estrogen modulates neuronal Bcl-xL expression and beta-amyloid-induced apoptosis: relevance to Alzheimer's disease. J Neurochem 72: 1552–1563

Pilgrim C (1994) Sex, hormones and the developing neuron. Eur J Histochem 38: 7–12

Pilgrim C, Hutchison JB (1994) Developmental regulation of sex differences in the brain: can the role of gonadal steroids be redefined? Neurosci 60: 843–855

Pitkänen A, Amaral DG (1993 a) Distribution of parvalbumin-immunoreactive cells and fibers in the monkey temporal lobe: the hippocampal formation. J Comp Neurol 331: 37–74

Pitkänen A, Amaral DG (1993 b) Distribution of parvalbumin-immunoreactive cells and fibers in the monkey temporal lobe: the amygdaloid complex. J Comp Neurol 331: 14–36

Planas B, Kolb PE, Raskind MA, Miller MA (1994) Galanin in the bed nucleus of the stria terminalis and medial amygdala of the rat: Lack of sexual dimorphism despite regulation of gene expression across puberty. Endocrinology 134: 1999–2004

Planas B, Kolb PE, Raskind MA, Miller MA (1995) Vasopressin and galanin mRNAs coexist in the nucleus of the horizontal diagonal band: a novel site of vasopressin gene expression. J Comp Neurol 361: 48–56

Polgar E, Antal M (1995) The colocalization of parvalbumin and calbindin-D28 k with GABA in the subnucleus caudalis of the rat spinal trigeminal nucleus. Exp Brain Res 103: 402–408

Raab H, Beyer C, Wozniak A, Hutchison JB, Pilgrim C, Reisert I (1995 a) Ontogeny of aromatase messenger ribonucleic acid and aromatase activity in the rat midbrain. Mol Brain Res 34: 333–336

Raab H, Pilgrim C, Reisert I (1995 b) Effects of sex and estrogen on tyrosine hydroxylase mRNA in cultured embryonic rat mesencephalon. Mol Brain Res 33: 157–164

Ragsdale Jr CW, Graybiel AM (1988) Fibers from the basolateral nucleus of the amygdala selectively innervate striosomes in the caudate nucleus of the cat. J Comp Neurol 269: 506–522

Reid S, Juraska J (1992) Sex differences in neuron number in the binocular area of the rat visual cortex. J Comp Neurol 321: 448–455

Reisert I, Pilgrim C (1991) Sexual differentiation of monoaminergic neurons-genetic or epigenetic? TINS 14: 468–473

Reisert I, Pilgrim C (1995) Catecholaminergic systems and the sexual differentiation of the brain. In: Segawa M, Nomura Y (eds) Age-related dopamine-dependent disorders. Monogr Neural Sci, Basel, Krager, 14: 216–224

Rhees RW, Shryne JE, Gorski RA (1990 a) Onset of the hormone-sensitive perinatal period for sexual differentiation of the sexually dimorphic nucleus of the preoptic area in female rats. J Neurobiol 21: 781–786

Rhees RW, Shryne JE, Gorski RA (1990 b) Termination of the hormone-sensitive perinatal period for differentiation of the sexually dimorphic nucleus of the preoptic area in male and female rats. Dev Brain Res 52: 17–23

Roberts GW, Woodhams PL, Polak JM, Crow TJ (1982) Distribution of neuropeptides in the limbic system of the rat: the amygdaloid complex. Neuroscience 7: 99–131

Romano GJ, Mobbs CV, Lauber A, Howells RD, Pfaff DW (1990) Differential regulation of proenkephalin gene expression by estrogen in the ventromedial hypothalamus of male and female rats: implications for the molecular basis of a sexually differentiated behavior. Brain Res 536: 63–68

Roos J, Roos M, Schaeffer C, Aron C (1988) Sexual differences in the development of accessory olfactory bulbs in the rat. J Comp Neurol 270: 121–131

Roselli CE, Resko JA (1993) Aromatase activity in the rat brain: hormonal regulation and sex differences. J Steroid Biochem Molec Biol 44: 499–508

Russchen FT (1982 a) Amygdalopetal projections in the cat. I. Cortical afferent connections. A study with retrograde and anterograde tracing techniques. J Comp Neurol 206: 159–179

Russchen FT (1982 b) Amygdalopetal projections in the cat. II. Subcortical afferent connections. A study with retrograde tracing techniques. J Comp Neurol 207: 157–176

Sar M, Lubhan DL, Franch FS, Wilson EM (1990) Immunohistochemical localization of the androgen receptors in rat and human tissues. Endocrinology 127: 3180–3186

Segovia S, Guillamón A (1993) Sexual dimorphism in the vomeronasal pathway and sex differences in reproductive behavior. Brain Res Rev 18: 51–74

Segovia S, Guillamón A (1996) Searching for sex differences in the vomeronasal pathway. Horm Behav 30:618–626

Shinoda K (1994) Brain aromatization and its associated structures. Endocrine J 41: 115–138

Shinoda K, Nagano M, Osawa Y (1994) Neuronal aromatase expression in preoptic, strial, and amygdaloid regions during late prenatal and early postnatal development in the rat. J Comp Neurol 343: 113–129

Shughrue P, Lane M, Merchenthaler I (1997 a) Comparative distribution of estrogen receptor-α and -α mRNA in the rat central nervous system. J Comp Neurol 388: 507–525

Shughrue P, Scrimo P, Lane M, Askew R, Merchenthaler I (1997 b) The distribution of estrogen receptor-α in forebrain regions of the estrogen receptor-α knockout mouse. Endocrinology 138: 5649–5652

Simerly RB, Chang C, Muramatsu M, Swanson LW (1990) Distribution of androgen and estrogen receptor mRNA-containing cells in the rat brain: an in situ hybridization study. J Comp Neurol 294: 76–95

Simerly RB, McCall LD, Watson SJ (1988) Distribution of opioid peptides in the preoptic region: immunohistochemical evidence for a steroid-sensitive enkephalin sexual dimorphism. J Comp Neurol 276: 442–459

Simerly RB, Swanson LW (1986) The organization of neural inputs to the medial preoptic nucleus of the rat. J Comp Neurol 246: 312–342

Simerly RB, Swanson LW (1988) Projections of the medial preoptic nucleus: a Phaseolus vulgaris leucoagglutinin anterograde tract-tracing study in the rat. J Comp Neurol 270: 209–242

Simerly RB, Swanson LW, Handa RJ, Gorski RA (1985) Influence of perinatal androgen on the sexually dimorphic distribution of tyrosine hydroxylase-immunoreactive cells and fibers in the anteroventral periventricular nucleus of the rat. Neuroendocrinol 40: 501–510

Simerly RB, Young BJ, Capozza MA, Swanson LW (1989) Estrogen differentially regulates neuropeptide gene expression in a sexually dimorphic olfactory pathway. Proc Natl Acad Sci USA 86: 4766–4770

Sorvari H, Soininen H, Piljarvi L, Karkola K, Pitkänen A (1995) Distribution of parvalbumin-immunoreactive cells and fibers in the human amygdaloid complex. J Comp Neurol 360: 185–212

Stefanova N (1998) GABA-immunoreactive neurons in the amygdala of the rat – sex differences and effect of early postnatal castration. Neurosci Lett. 255: 175–177

Stefanova N, Bozhilova-Pastirova A, Ovtscharoff W (1996) Influence of sex on brain structure. Mol Medicine, 1 (3–4): 60–64

Stefanova N, Bozhilova-Pastirova A, Ovtscharoff W (1997a) Distribution of GABA-immunoreactive nerve cells in the bed nucleus of the stria terminalis in male and female rats. Eur. J. Histochem. 41: 23–28

Stefanova N, Bozhilova-Pastirova A, Ovtscharoff W (1997b) Sex differences of parvalbumin-immunoreactive neurons in some rat brain areas. Biomed. Rev. 7: 91–96

Stefanova N, Bozhilova-Pastirova A, Ovtscharoff W (1997c) Developmental changes in the pattern of parvalbumin-immunoreactivity distribution in the lateral amygdaloid nucleus of the rat. Ann. Anat. 179: p. 195

Stefanova N, Bozhilova-Pastirova A, Ovtscharoff W (1998) Sex and age differences of neurons expressing GABA-immunoreactivity in the rat bed nucleus of the stria terminalis. Int. J.Dev. Neurosci. 16: 443–448

Stoykov I, Kolev J, Karamfiloff K, Baeva N, Stefanova N, Bozhilova-Pastirova A, Ovtscharoff W (1998) Sex differentiation in the rat striatum. immunocytochemical study of parvalbumin-containing neurons. Symposium of clinical anatomy, Varna, Scripta Sci. Med., Γ, Suppl., p. 41

Swaab DF, Fliers E, Partiman T (1985) The suprachiasmatic nucleus of the human brain in relation to sex, age and dementia. Brain Res 342: 37–44

Swaab DF, Hofman MA (1988) Sexual differentiation of the human hypothalamus: ontogeny of the sexually dimorphic nucleus of the preoptic area. Dev Brain Res 44: 314–318

Szot P, Dorsa DM (1993) Differential timing and sexual dimorphism in the expression of the vasopressin gene in the developing rat brain. Dev Brain Res 73: 177–183

Torii M, Kubo K, Sasaki T (1996) Influence of opioid peptides on the priming action of estrogen on lordosis in ovariectomized rats. Neurosci Lett 212: 68–70

Urban JH, Miller MA, Drake CT, Dorsa DM (1990) Detection of vasopressin mRNA in cells of the medial amygdala but not the locus ceruleus by in situ hybridization. J Chem Neuroanat 3: 277–283

Usunoff K, Romansky K, Blagov Z, Bezlov S, Penev L (1979) Efferent projections of corpus amygdaloideum to the brain stem. Neurol Psych Neurosurg, 6: 394–403

Van Furth WR, Wolternik G, van Ree JM (1995) Regulation of masculine sexual behavior; involvement of brain opioids and dopamine. Brain Res Rev 21: 162–184

Van Leeuwen FW, Caffe AR, De Vries GJ (1985) Vasopressin cells in the bed nucleus of the stria terminalis of the rat; sex differences and the influence of androgens. Brain Res, 325: 391–394

Van Ziegler NI, Schlumpf M, Lichtensteiger W (1991) Prenatal nicotine exposure selectively affects perinatal aromatase activity and fetal adrenal function in male rats. Dev Brain Res 62: 23–31

Vankova M (1991) Immunohistochemical characteristic of the central amygdaloid nucleus, the interstitial nucleus of the stria terminalis and their projections. PhD Thesis, Sofia

Vito CC, Fox TO (1982) Androgen and estrogen receptors in embryonic and neonatal rat brain. Dev Brain Res 2: 97–110

Wagner CK, Morrell JI (1997) Neuroanatomical distribution of aromatase mRNA in the rat brain: inductions of regional regulation. J Steroid Biochem Mol Biol 61: 307–314

Wang Z, De Vries GJ (1995) Androgen and estrogen effects on vasopressin messenger RNA expression in the medial amygdaloid nucleus in male and female rats. J Neuroendocrinol 7: 827–831

Weisz J, Brown BL, Ward LL (1982) Maternal stress decreases steroid aromatase activity in brains of male and female rat fetuses. Neuroendocrinol. 35: 374–379

Weisz J, Ward IL (1980) Plasma testosterone and progesterone titres of pregnant rats, their male and female fetuses, and neonatal offspring. Endocrinology 106:306–316

Weller KL, Smith DA (1982) Afferent connections to the bed nucleus of the stria terminalis. Brain Res 232: 255–270

Wimer RE, Wimer C (1985) Three sex dimorphisms in the granule cell layer of the hippocampus in house mouse. Brain Res 328: 105–109

Witelson SF, Glezer II, Kigar DL (1995) Women have greater density of neurons in posterior temporal cortex. J Neurosci 15: 3418–3428

Wood RI, Newman SW (1995) The medial amygdaloid nucleus and medial preoptic area mediate steroidal control of sexual behavior in the male Syrian hamster. Horm Behav 29: 338–353

Woodhams PL, Roberts GW, Polak JM, Crow TJ (1983) Distribution of neuropeptides in the limbic system of the rat: the bed nucleus of the stria terminalis, septum and preoptic area. Neuroscience 8: 677–703

Wouterlood FG, Hartig W, Bruckner G, Witter MP (1995) Parvalbumin-immunoreactive neurons in the entorhinal cortex of the rat: localization, morphology, connectivity and ultrastructure. J Neurocytol 24: 135–153

Wright CI, Beijer AVJ, Groenewegen HJ (1996) Basal amygdaloid complex afferents to the rat nucleus accumbens are compartmentally organized. J Neurosci 16: 1877–1893

Yamano M, Hillyard CJ, Girgis S, Emson PC, MacIntyre I, Tohyama M (1988) Projection of neurotensin-like immunoreactive neurons from the lateral parabrachial area to the central amygdaloid nucleus of the rat with reference to the coexistence with calcitonin gene-related peptide. Exp Brain Res 71: 603–610

Yokosuka M, Okamura H, Hayashi S (1995) Transient expression of estrogen receptor-immunoreactivity (ER-IR) in the layer V of the developing rat cerebral cortex. Dev Brain Res 84: 99–108

Young MW (1936) The nuclear pattern and fiber connections of the non-cortical centers of the telencephalon of the rabbit, (*Lepus coniculus*). J Comp Neurol 65: 295–401

Yuri K, Kawata M (1994 a) Estrogen affects calcitonin gene-related peptide- and methionine-enkephaline-immunoreactive neurons in the female rat preoptic area. Neurosci Lett 169: 5–8

Yuri K, Kawata M (1994 b) Region-specific changes of tyrosine hydroxylase-immunoreactivity by estrogen treatment in female rat hypothalamus. Brain Res 645: 278–284

Zhou J-N, Hofman MA, Gooren LJG, Swaab DF (1995) A sex difference in the human brain and its relation to transsexuality. Nature 378: 68–70

Subjext Index